THE UNCONSTRUCTABLE EARTH

MEANING SYSTEMS

THE UNCONSTRUCTABLE EARTH

An Ecology of Separation

FRÉDÉRIC NEYRAT
TRANSLATED BY DREW S. BURK

Fordham University Press: New York 2019

This work received the French Voices Award for excellence in publication and translation. French Voices is a program created and funded by the French Embassy in the United States and FACE (French American Cultural Exchange). French Voices logo designed by Serge Bloch.

This book was originally published in French as Frédéric Neyrat, *La part inconstructible de la terre: Critique du géo-constructivisme*, Copyright © Éditions du Seuil, 2016.

Cet ouvrage a bénéficié du soutien des Programmes d'aide à la publication de l'Institut Français.

This work, published as part of a program of aid for publication, received support from the Institut Français.

Fordham University Press gratefully acknowledges financial assistance and support provided for the publication of this book by the University of Wisconsin-Madison.

An advance excerpt from Chapter 13 of this book appears in *diacritics* 45, no. 3 (2017).

Library of Congress Cataloging-in-Publication Data available online at https://catalog.loc.gov.

Printed in the United States of America

21 20 19   5 4 3 2 1

First edition

for my friends who continue to go on living

CONTENTS

INTRODUCTION

Reconstructing the Earth?

> The combined circumstances that we live on Earth and are able to see the stars—that the conditions necessary for life do not exclude those necessary for vision, or vice versa—is a remarkably improbable one.
> —Hans Blumenberg, *The Genesis of the Copernican World*

> You did not wish to see the face of the Unknown; you will see its mask.
> —Victor Hugo, *Préface de mes œuvres et post-scriptum de ma vie*

A NEW GRAND NARRATIVE

Humanity's conquest of outer space is over? No. We have simply discovered a new planet: Earth. An Earth we would supposedly be able to reconstruct and pilot thanks to the exploits of an absolute engineering process. Geo-constructivism will be the name bestowed upon this voracious discourse that considers the Earth as the consenting prey of an integral conquest. Aggregating a diverse number of philosophical, economic, and scientific lines of inquiry, geo-constructivism asserts itself at the intersection of several discourses: the discourses of engineers and architects who would like to transform the Earth into a pilotable machine; biologists who would rather spend their time resurrecting already extinct species rather than protecting those that are still alive; political strategists offering solutions for global governance; the advent of new markets by businessmen who view climate change as a new industry for economic speculation; geographers enthralled by the power of humanity within the age of the Anthropocene; sociologists and anthropologists proclaiming that there is no common world and so it is up to us to build one; essayists promoting nuclear energy for all; prophets declaring the death of nature or the birth of the transhuman; philosophers inviting us to accelerate our technological control over existing society; paradoxical ecologists simultaneously lauding the merits of fracking and dreaming of the disappearance of any form of ecology containing a political dimension. How did such a discourse emerge? How can we begin to explain its ever-increasing influence and irresistible hegemony? Who are its principal spokespersons? And what does its future have in store?

Geo-constructivism is not some definitively unified theory. As with any discourse with a hegemonic vocation, geo-constructivism contains contradictions, vanishing points, and ambiguities. In order to untangle this internal heterogeneity, the present book begins by proposing an empirical philosophical inquiry into the concept of the Anthropocene in its relation to recent solar engineering projects (the invention of a chemical shield to protect the Earth from solar radiation), synthetic biology (the advent of organisms with new or improved traits), and finally "Earth stewardship." Next, we will decipher the form of ecological thought compatible with these projects, before then proposing to conclude our inquiry with an alternative conception of nature and our relation to the Earth. This progressive passage from the empirical to the speculative will allow us to analyze, under different aspects, the fundamental fantasy at the heart of geo-constructivism: claiming that the Earth, and everything contained on it—ecosystems and organisms—humans and nonhumans, *can* and *must* be reconstructed and entirely remade.

If the Earth *can* indeed be reconstructed, this is only possible because of the geo-constructivists' view that nature—as an independent entity and force—has been overtaken by the techno-industrial power of humanity; if the Earth *must* be reconstructed, this is because it would be the only solution to solving ongoing environmental problems. The project for a "general reconstruction of the world" would be a project of *general ecological interest*.[1] Mark Lynas provides a perfect summary of this doctrine in his book *The God Species*: *Saving the Planet in the Age of Humans*. Lynas writes, "Nature no longer runs the Earth. We do. It's our choice what happens from here."[2] The thesis of our present work is that, as fantastical as the doctrine of geo-constructivism may sound, it is in the midst of ushering in a new grand narrative, the new myth of our current age: not a mere fiction without any true repercussions but a discourse capable of legitimizing real economic decisions, social practices, lifestyles, laws, institutions, and the guiding orientations for civilization. What does the geo-constructivist grand narrative have to tell us? What promise does it have in store for us? The narrative unabashedly declares: Yes, it's clear, the world is in the grips of ecological disasters, but we mustn't forget that these disasters were caused by humanity and not by some sort of obscure fate; far from extracting a sense of guilt from this current situation, we must on the contrary, *extract a profit* from this telluric power. If we have *damaged* the Earth, it's because we had the power at our disposal to do so. We have made the Earth poorly? Then we must repair it, reprogram it—*reconstruct it!*

TECHNOLOGY OF A WORLD WITHOUT NATURE

Geo-constructivism's exploit: the capability of recycling the project of scientific modernity consisting of becoming "masters and possessors of nature" (Descartes), while simultaneously solving the environmental disasters intrinsically associated with this same conquest. In order to return to the concepts proposed by Ulrich Beck, geo-constructivism is presented as a reflexive discourse that, having analyzed and overcome the errors of the first modernity (founded on the idea of progress), would have learned how to take into consideration the criticism of the ecologists as well as the risks generated by industrial technologies.[3] However, the dream of an "absolutely modern society" has instead turned into an utter nightmare.[4] In stark contrast to Beck's hopes, geo-constructivism's reflexivity was not born out of a true environmental critique of modernity but rather reinforced modernity's original project without improving on it, without purging it of its original flaw—its anthropocentric drive toward conquest, whether in the form of technology (the control of a nature *out of joint* with its power) or culture (the human all too human decision, relative to the contingent split between the artificial and the natural)—namely, geo-constructivism's inability to open itself up to a consideration of the entire universe. In this sense, the geo-constructivists are *hypermodern*, more modern than Bacon or Descartes. Indeed:

1. The geo-constructivists explicitly integrate techno-industrial hazards. As Günther Anders already exclaimed in 1966, we don't realize we are mere sorcerer-apprentices producing destruction;[5] from now on, this knowledge has been integrated as simply one tiny element of a supposedly superior knowledge: as if by magic, the sorcerers will have progressed from the stage of apprentices to that of masters.

2. The geo-constructivists are not striving to conjure these dangers by way of the self-limitations of industrial and technological power but rather by way of an increase in anthropogenic modification. The geo-constructivist knows he is playing with possibles and that his intervention will provoke unexpected outcomes; he is the sorcerer-apprentice wielding a magic wand with the hand of a master, but this merely confirms his appetite for transformation. To paraphrase Hölderlin, geo-constructivist belief can be summarized thusly: Whereas one can detect a danger looming within the industry of nanoparticles, the fields of genetically modified crops and foods, nuclear energy, synthetic biology, and the artificial modification of the climate, one can also detect

a saving grace.[6] In this sense, the fundamental promise of geo-constructivism is not progress (like it was for the Saint-Simonians in the nineteenth century) but the mere survival of humanity: From now on, progress is a *secondary benefit* of a planetary lifesaving program.

Let us be very clear right from the outset: This book is not a general denunciation of technology in the name of some kind of return to a pure form of nature. Our perspective consists in analyzing the manner in which the conception of nature—which is at the heart of the geo-constructivist program—and its hypermodern horizon, is intimately correlated to the technological possibility it is striving to actualize. This correlation could be stated in the following way: The geo-constructivist program consists of privileging technologies that consider nature as nonexistent. The geo-constructivists embrace statements such as "the conditions for human life were not natural and never have been" or "the idea that humans must live within the framework of the natural environmental limits set by our planet denies the reality of our entire history." The cold and inevitable conclusion drawn by Erle Ellis, a geography professor at the University of Maryland, is harsh, and perfectly geo-constructivist: "The environment will be what we make of it."[7] From now on, nature is posited as a "nonbeing" and incapable of disturbing the anthropogenic modification of environments in any way whatsoever. The main thesis of geo-constructivism is therefore not "naturalist," that is, as Philippe Descola claims in his anthropological work—founded on a dualism of humans and nonhumans—but rather *anaturalist*: without nature, ignorant of it.[8] Geo-constructivism's anaturalist optic does not *tend* toward *separating* two worlds (into the world of humans and the world of nature) but completely erases and denies the existence of one of these worlds. And yet, a denial is not a separation, which recognizes an inherent difference between the two terms. A denial is a refusal to recognize the existence of one of these terms for the benefit of the lone identity of the other term. For the promoters of anaturalism, the only world that exists—the only world that must exist—is technology: You will read a number of writings about the "end of nature," but anyone who would proclaim the end of technology would immediately be locked up in a psychiatric hospital.

We could, of course, follow in the footsteps of Paul Feyerabend and choose to revert our gaze all the way back to Parmenides for a philosophical origin of the *denaturalization of nature*: For Parmenides, nature as abstract, homogenous, and distant from any kind of experience would seem to already expel anything that would firmly bind it to the world of the living.[9] Nevertheless, nature remains a model for the paradigm of antiquity, as this well-known statement by Aristotle attests to: "Art (techné) on the one hand completes to a certain degree what nature is incapable of effectuating, and on the other hand, it imitates it."[10] However, by substituting God for nature, the monotheisms will cleverly devalue the latter, removing its power and thereby preparing the terrain for the new science of the seventeenth century: Nature becomes an inanimate and mathematizable material, upon which a human fashioning is applied that always tends toward extending the limits of the possible and transforming the impossible into the possible. From then on, anaturalism assures its hold over the Earth. With the arrival of a geo-constructivist hypermodernity, it is precisely *the idea of nature itself* that disappears within the aftermath of the substitution of nature by artificial entities whose objective is to integrate, digest, and reprogram all natural alterity. Nature becomes "biodiversity," "preservation of social services" (the water supply, pollination, etc.), "resources"—merchandise.[11] From then on, anaturalism clearly appears as the condition for the ontological possibility of technologies whose goal is to replace nature. The goal of these technologies of substitution is not simply to conquer nature but to *remake* it, by substituting its own technological power for nature.

We will insist on using this verb: to *reconstruct* [refaire] can also mean to *modify* (such as when one refers to genetically modified organisms—GMOs); to find a *substitute* (as in the case of an artificial uterus);[12] or if we consider the idea that the human form of life and its human body are obsolete and must therefore be replaced by "transhumans" or "posthumans." To modify, substitute, and replace are all at work within the manner in which geo-constructivism envisions remaking the Earth. In order to understand the way in which geo-constructivism has translated, for its own sake, the *passion of remaking* and thus prolonging the conquest of the modern, let's read what Lewis Mumford already proclaimed in 1966:

> Our age is passing from the primeval state of man marked by his invention of tools and weapons for the purpose of achieving mastery over the forces of nature, to a radically different condition, in which he will have not only conquered nature, but detached himself *as far as possible* [my emphasis] from the organic habitat. With this new "megatechnics," the dominant minority will create a uniform, all-enveloping, super-planetary structure designed for automatic operation.[13]

Here, we can slowly begin to see a strange topology unfolding: It's as if the geo-constructivists view themselves as residing off-planet, outside the Earth, without any kind of vital relation with the ecosphere, detached and separated *as far away as possible* from the Earth object to be reformatted. It's from this seemingly extraplanetary position that the geo-constructivists reflect on how to produce a "super-planetary structure" for the happy select few who would supposedly be capable of controlling it. Remaking the Earth would thus actively allow for the realization of Teilhard de Chardin's fantasy that he prophesized as our unavoidable fate: the passage from the "biosphere to the noosphere," "planetized humanity," that is to say, to the sphere of the human mind that, thanks to its technological globalization, would supposedly be capable of materially and spiritually detaching itself from an organic form of life judged to be outmoded at an evolutionary scale.[14]

If our analysis is correct, it follows that the contemporary thought regarding ecology, and the anthropological approach accompanying it, must refrain from endlessly questioning itself about the dualisms of nature/technology or nature/culture, those great divides attributed to the modernists of the West as well as the means for going beyond these divisions thanks to hybridizations, more transformations or more interactions. It's not a question of "loosening the vice grip" of dualism (Catherine Larrère and Raphaël Larrère) but of contesting this category.[15] To speak of dualism is to incite the belief that there is something called two [il y a du deux] when what is at stake is the disappearance of one of these terms named within the supposed divide. To believe in something called the two when there isn't such a thing—to believe that there is that which is *wild* and that which is artificial when the former is disappearing under a deluge of concrete[16]—is to play the game of the One, that is to say, the game of domination. In the end, this is not simply a question for ethics or philosophy but, above all, for politics: Is it possible for someone to forgo participating within the extension of an empire of a world without nature, to remain outside this enterprise that is attempting an intellectual and material eradication of all alterity from the human experience? *Instead of attempting to transcend the so-called dualisms and Great divides (between nature and culture, humans and nonhumans, etc.), is it possible to look for what stands behind the Great denial (of alterity)?* As Eduardo Viveiros de Castro notes, "because the Europeans thought that America was a world without humans, the Indians became humans without a world": Their world was destroyed by the Moderns.[17] But with the arrival of geo-constructivist hypermodernity, this great denial of alterity has become a planetary religion and the constructivist approach of a "worldless humanity" can end up being applied to *everyone.*[18] From this point on, escaping from the great denial of alterity becomes a major concern for contemporary political ecology.

REFORMATTING THE EARTH: PILOTING A MANAGEMENT MACHINE

In order to better understand the nature-technology apparatus associated with the geo-constructivist program, the first part of our book will begin by taking an interest in one of its most emblematic projects—geoengineering, and more specifically climate engineering—in other words, the attempt to control the climate by way of the mediation of its technological optimization. A hyper-Cartesian apparatus of mastery and possession consisting of remaking terrestrial nature, climate engineering is the mirror in which the Anthropocene would like to see its own reflection—an era wherein they tell us (and repeat it until they're blue in the face) that the human species has become a force to be reckoned with—oh how wonderful! We've become a "major geological force." We will also study the links connecting the dominant discourse of the Anthropocene with geoengineering. For it's not by chance if, in 2002, the co-inventor of the term *Anthropocene*, Paul Crutzen, mentions in an article the possibility of "geoengineering projects on a large scale" with the end goal of artificially "optimizing" the climate—an article where he also has no reservations referencing Teilhard de Chardin.[19] This is the same Paul Crutzen who proposed, in a rather bombastic article from 2006, to send tons of sulfur dioxide into the atmosphere in order to create a chemical "shield" that would purportedly be capable of protecting us from the sun and that would consequently recool the planet.[20]

One mustn't be fooled into believing this is a case of mere science fiction masquerading as science fact that would only garner interest from a few fringe scientists or fans of Hollywood films. In order to be convinced of this, all one has to do is review the increase in funding for such projects and the interest that President Obama and an influential newspaper such as the Wall Street Journal have in questions concerning geoengineering. We should also consider the increasing number of conferences, academic articles, and books for the general public devoted to the question of geoengineering.[21] Even the IPCC, in one of the last paragraphs from its 2013 "Summary for Policymakers," finally mentions the concept of "geoengineering."[22] Certain members of well-known ecological associations, such as WWF, no longer deny the "possibility" of "geoengineering."[23] This "possibility" no doubt explains why certain *think tanks* that once simply chose to deny climate change now appear to have finally come to accept it. Are we witnessing the arrival of a new market ripe for conquest?[24] A new frontier for capitalism? Of course, the risks and potential dangers of such projects are often evoked as well; in line with the American Academy of Sciences, we often prefer to speak of "climate interventions" rather than geoengineering (a term that is too often associated with trying to control climate, a task deemed beyond our grasp).[25] But in no way do these potential risks prevent more and more ongoing studies and research regarding the national and international rules and regulations that would—eventually—be the guiding framework for this geo-technology.[26] As a technological object that is also juridical, economical, social, and environmental, the "climate bubble" is the so-called "plan B" that many would like to see turn into "plan A" in order to fight climate change. A "plan B" whose dangers are well known to its promoters, but such is the new doctrine of the geo-constructivists: not some utopia of a constant progress but the pragmatism of a lifesaving expedition of the Earth that will require taking risks.

We will show that the representation of the Earth preferred by the geo-constructivists is that of an image of a *hollowed-out box* one can reformat at will. The dynamisms of terrestrial nature are only recognized as existing when it is necessary to legitimize a reformatting project: One considers a dynamic as a biospheric process that one prepares for remodeling, skillfully managing according to the economic imperatives of the moment. In other words, natural dynamics become the arguments for turning the Earth into a computer. Inside his own kind of fantasy space, the geo-constructivist sees himself as some kind of rogue off-planet agent, with demiurgic powers for reshaping the Earth from the outside. And in this regard, the geo-constructivist embodies the dream of Richard Buckminster Fuller—the famous architect-designer and American inventor active from the 1950s through the 1970s—known for baptizing the Earth as a "spaceship." "I've often heard the question, 'I wonder what it would be like to be on board a spaceship' and the answer is very simple. What does it *feel* like? We are all astronauts."[27] And yet, this metaphor created a double meaning: It's not simply that the inhabitants of Earth are viewed as extraterrestrials but that the Earth is no longer viewed as the cradle of humanity and is instead envisioned as a kind of exoplanet that we must "terraform." A concept plucked right out of science fiction literature, the idea of "terraforming" first of all signifies the process of deliberately modifying another planet in order to render it similar to Earth and, as a consequence, making it inhabitable for human beings. But from now on, it's the Earth itself that we would like to terraform at our own convenience: The Anthropocene inherited its imaginary from the era of space exploration and the era's ambitions for extrasolar colonization.[28] In fact, it's as if everything was set up for the end of the Space Age to coincide with the promotion of the Age of Man. Of course, today we're talking about "plan C," which would consist of creating a "living spaceship" capable of extracting humanity from a dying planet and shepherding it to a new, more welcoming one.[29] But this is not the imaginary of some sort of virile colonization project but rather, and this is the moment to put it out there, an imaginary of *revival*—of the 1950s, a speculation concerning the various alternative days to come. The eventual colonization of the Moon or Mars having been abandoned once and for all, or at the very least put on long-term hold, it's now the Earth itself that has become the object of a technological colonization project. From now on, the frontier of capitalism is no longer some kind of dreamed-of *beyond* whose first attempts were initiated by Sputnik and the Apollo missions. Capitalism's new frontier has now been left in the hands of the geo-constructivists and their bio-constructivist allies: Commodify the atmosphere and securitize the planet (green finance), artificialize and manage the planetary climate (synthetic biology), such are the fronts put forth regarding the *shifting of the frontier* toward the body of the Earth.

After having deciphered the imaginary foundation of the project and the geo-constructivist apparatuses, it will then be possible to understand the precise function of the concepts of "piloting" or "Earth stewardship." A notion proliferating within environmental and scientific literature, Earth stewardship situates science as the discipline that must facilitate the guided management of socioecological changes in view of the well-being of humans and their inherent resilience. Distinct in its appearance as a kind of geo-constructivist voluntarism, the concept of management (of biodiversity, of environments, etc.) leaves one to believe humans would do nothing more than simply steer or manage various dynamisms without truly being capable of controlling them. By way of their seemingly *soft* appearance, being cooperative and bearers of a morality of respect, the concepts of management and stewardship have the unfortunate tendency of making us forget about the technologies that make them possible: Indeed, what else would this stewardship be that would be applied to environments modified by climate engineering, environments that have their sights set on human well-being if not the *self-service* of the age of humans? Furthermore, the concept of management was first applied to a machine: the piloting of a machine that we could call the management machine—in other words, the social, economic, political, and technical megamachine underlying all stewardship and governance of the ecosphere. From now on, we can define the Anthropocene as a grand narrative seeking legitimization for the installation of a global, pilotable, management machine: Its politics propped up by the powers of engineers, and its fantasy being the possibility of an integral project of terraforming.

THE THEORETICAL FOUNDATIONS OF ECO-CONSTRUCTIVISM

This will lead us to understand why certain *think tanks* and businesses in the United States simultaneously refuse to recognize the anthropic origins of climate change *and yet at the same time* are also calling for the geoengineering of the climate. It's true, geoengineering is presented as a means for aiding the continual tempered burning of fossil fuels and, therefore, as a means to help maintain the same kind of development, society, and domination organized and hoped for by the militants of the geo-constructivist program.[30] But how is it that a large number of supposed ecologists have married themselves to these ideas? The second part of the present book will attempt to demonstrate the following thesis: Ecological and environmental thinking today is dominated by an eco-constructivist current that is less and less apt at contesting geoengineering and the anaturalist position and is more and more opposed to any truly antiestablishment political ecology. To say that this eco-constructivism, elaborated within the humanities and social sciences as well as scientific ecology, is becoming more and more compatible with the geo-constructivist project in no way implies that ecological or environmental thinking has always been eco-constructivist, nor does it mean that any kind of ecological thinking or any environmental philosophy is from now on constructivist. We want to simply show to what extent eco-constructivism has begun to shake up the ongoing debates—from green Marxism to environmental ethics, passing by thoughts concerning Gaia and the pragmatism of the composition of worlds—and in this sense, eco-constructivism has indeed imposed its hegemony. Identifying the dominant thought of ecology under the name of eco-constructivism is the means by which we will seek to establish a new political ecology that is resolutely anticonstructivist.

Eco-constructivism is located at the intersection of several different lines of thinking.

A. From a historical point of view, the roots of eco-constructivism are to be found in the ecology of resilience. Born in the 1970s, this theory is based on an affirmation of the ontological instability of ecosystems and the rejection of any idea of nature being balanced or stationary.

B. The tree trunk of eco-constructivism is composed of approaches defended by the ecomodernists, these post-environmentalists who proclaim the death of environmentalism,[31] castigating any desire for protecting spaces and forms of life from the conquest of capital growth and profit, rejecting any idea of placing limits on alterations and anthropogenic artificializations.[32] In light of this, we must show that the so-called "nonmodern" constructivists and sometimes even certain "pragmatic" ecologists are often in agreement with the fundamental ontological options of the ecomodernists.[33]

C. Finally, eco-constructivism is clearly raised to the level of its paroxysm and its greatest truth by certain ungrounded surgeons such as the trendy accelerationists who, in recognizing the pertinence of the stakes of climate change, call for a "promethean politics of maximal control over a society and its environment,"[34] as well as certain transhumanist currents completely aware of the possibility of the risks of the extinction of the human species, and who dream of posthuman forms of life. A root, a tree trunk, strange growths—but, from a conceptual viewpoint, what are the common points or, rather, the common *fields of forces* that inform the eco-constructivist domain?

1. First and foremost, *a euphoric proclamation of the death of nature*. This is one of the most distinctive traits of eco-constructivism and—above and beyond

the question of ecology—all forms of constructivism: There is no nature. For the eco-constructivists, nothing is given, everything is a process, and in this sense, *everything can be constructed*. The inevitable consequence: *The world is there to be endlessly made and remade.* Far from being considered as bad news, the death of nature is celebrated to the extent that it can be used to fight against the idiocy of the supposed sad ideas of a natural limit or finitude that the environmentalist movement desperately attempts to defend. Swept up by its anaturalist passion, eco-constructivism is unable to see that the murder of nature is as old as the so-called "West."

2. Next, the awareness of "uncertainty" as the new hidden god, the *deus absconditus* of the risk society. In order to clarify this point, we will carefully examine the concept of resilience and how it was forged in the 1970s. The fundamental idea of what we will call the ecology of resilience is the following: Far from being a curse, the ontological instability of ecosystems, the "turbulence" of the world, is the reservoir that organisms and societies can tap into in order to adapt themselves to new situations, to transform themselves, and thus survive.[35] It is, of course, important to recognize the real contributions of theories that insist on this ontological precarity of the world. Ilya Prigogine and Isabelle Stengers[36]—and Alfred North Whitehead before them[37]—have maintained that Western tradition has strived to conjure chaos, disorder, and imbalance and to place them onto a second plane. In this respect, the rehabilitation of philosophies of unstable flows during the 1970s (Michel Serres, Gilles Deleuze, Luce Irigaray) was a saving grace. Nevertheless, this ontological thesis has been used by eco-constructivism as an a-critical justification of the world as such: Everything changes, so why should we be opposed to any economic changes at all? Everything is unstable, so why should we demand *social security* or any kind of insurance from the State? Within such a theoretical framework, resilience is nothing more than that which makes humans change in order to better adapt *by force* to economic, social, and ecological disasters without ever seeking to get at the heart of the primary causes of these disasters. By this same token, a British study from 2011 entitled *Migration and Global Environmental Change* valorizes migrations not as a kind of misfortune but as a way of increasing the capacity of individuals (and not of collectives) for adapting to environmental disaster, even their capacity for anticipating it; these studies tell us that it's better to migrate before the disaster . . .[38] We can immediately see how such an idea is in line with climate engineering, which is a way of responding to the climatic symptom without ever having to get at the exact *causes*—be them social, political, or economic—of these changes.[39] The turbulence of the anthropogenic sky has replaced the unfathomable fate that the Sky Gods once inflicted upon humanity.

3. Then there is the idea, repeated as a mantra, according to which everything is interconnected. This is what we will call *the principle of the principles* of ecology and environmentalism. From Ernst Haeckel to James Lovelock and Isabelle Stengers passing by John Muir, Aldo Leopold, Barry Commoner, and Arne Næss, ecological thinking in all its variations was largely and legitimately founded upon this principle with its sights set on fighting the denial of relation—between humans and their environment, reason and the body, the industrial revolution and its so-called collateral damage—on top of which the modern era laid its foundation. But, assisted by the teletechnologies that encircle the globe, today the notion of interconnection constitutes a kind of theoretical, economic, and political trap that reveals itself in the following way: Since everything is a network, linked together, interconnected, then there can be no distance possible in relation to the world in which we live, so we should simply accept the world as it is, with its ontological turbulence; be pragmatic and take care of one's resilience. *Interconnection and uncertainty are two sides of the same approach to the world*: Since everything is connected, we never know to what extent a phenomenon can be propagated; since the borders of the world are uncertain, the relation between things is more consistent—more permeable, more contagious—than we would have ever thought.

4. And lastly, *an unwavering faith in technological modernity*. The eco-constructivists have a firm belief in modernity and its technological virtues. Of course, it should be understood that the modernity of the "ecomodernists" is not of the highest degree. It claims itself to be "reflexive," and the progress it promotes has become sensitive to the "risks" and "unexpected consequences" that are always a part of an "uncertain world." Nevertheless, in no way does this prevent the eco-constructivists from sharing, along with the geo-constructivists, the same passion for the latest inventions of the capitalist techno-industry. More specifically, we will engage with the theory put forth by Bruno Latour, to the extent that Latour is one of the major figures of eco-constructivism and in light of his notable influence in regard to the institutional edification of the discourse that has today become hegemonic concerning ecology. It's true, Latour claims that he is not modern; however, we will attempt to show that his relationship with technologies and his anaturalist position make him into a profoundly modernist thinker. We must once again clarify what we mean by this term *modernist*: Descartes, Galileo, and Bacon constitute scientific modernity starting from a methodological rupture with forms of knowledge that were retroactively deemed premodern, but this modernity is in continuity with the *phantasmatic* desires of the premodernists. *Modern is the one who seeks out the elixir of long life with the science of Descartes*. By erasing the methodological break between premodern/modern, all the while

maintaining a steadfast faith in the technological elixir, the sociology of the "composition" of worlds has let a number of premodern, para-alchemical flows of affect wash over contemporary theory, where nonhuman and human "assemblages" task themselves with singing the praises of global technicity, of the capitalism that underpins it, and the consumerism that accompanies it. Indeed, it's not enough to simply explain that phenomena are hybrid in order to contest anthropocentric modernity; one must also show that these hybridizations are not simply the effect of human practices and technologies. With Latour and his post-environmentalist allies, we are condemned to reinforcing modernity and accentuating it all the way to its hypermodern limit. In this light, *a hypermodernist is the one who thinks that the extension of life has less to do with humans than their denaturalized successors*—some sort of "planetized" humanity, or some kind of *"Anthropos"* plunged "into an epoch that is at once post-natural, post-human, and post-epistemological."[40]

We can see quite well how these four points, these four fields of forces that we have just described, conspire in order to form a consistent discourse: There is no nature; nothing is given, nothing can assure us of any initial rule governing things; namely, everything is ontologically turbulent, full of surprises, and politically uncertain; in this sense, we can immediately get rid of the idea of a limit or separation as well as any principle that could stabilize a behavior in advance;[41] we must accept the technologies propelled by the dominant economy, in other words "we must love our monsters" (Latour[42]); and, in a pragmatic fashion, we must task ourselves with observing the effects of these technologies. This present book will make the case against this technological and political conformity and herd mentality in favor of another political and theoretical ecology: an *ecology of separation*. We shall now attempt to clarify this expression.

FROM THE PARADOX OF THE ANTHROPOCENE TO AN ECOLOGY OF SEPARATION

Geo-constructivism and eco-constructivism seem to be opposed to each other in one very crucial aspect: Whereas geo-constructivism perceives of itself as fantastically projected into the stratosphere, confronted with an Earth object that it attempts to control, eco-constructivists strongly insist on "generalized" interconnection, "entanglements" (Stengers), and "attachments" (Latour) that connect *all* beings, including humans. Because of this, we will find in the same book, and from the same pen, both the recognition of the absolute interconnections that bind us to the ecosphere and, at the same time, the affirmation of a dominion of human sovereignty where "Man," the "human species," "humanity," "anthropos," or (human) "culture" affirms its right to the recon-struction of the ecosphere, and even the entire planet. Such is the paradox of the Anthropocene that haunts the contemporary thought of ecology: Hu-manity is at once a) a morphogeologically superpowerful subject, exterior to body-objects (the body of animals, humans, or the Earth) that it declares ca-pable of controlling, managing, etc., and b) that which, on Earth, suffers the consequences of its actions to the extent that humanity is linked to the forms of life and abiotic elements that constitute the planet. Like a Prometheus bound to some sort of cyber-eagle, humanity dreams of a solar shield but doesn't yet know what kind of thermic protection it would be capable of inventing in order to protect itself from this scorched shield.[43]

In order to escape from this paradox of the Anthropocene by another way than the hypermodern path—the growth of high-tech modernity by which we will strive to control the negative effects of our lack of mastery of the planet—this present work proposes another approach for ecology: an *ecology of separa-tion*. Some would perhaps see here another paradox: If ecology is a thought of relations, then how can it promote separation? Is this not precisely the Car-tesian method and its geo-technological extension? And yet, we must come to terms with the fact that in order to exit the paradox of the Anthropocene, ecological thinking must fight on two fronts, not simply one:

1. Against the division of the Humanity-subject/Earth-object, we must truly propose a set of relations and demonstrate to what degree nothing is truly isolated: A worldless-humanity is nothing more than an ontological and envi-ronmental impossibility destined toward self-destruction.

2. But contrary to the madness of generalized interconnection, what we truly require is a distancing.[44] An ecology of separation strives to reconnect that which has been *split* into two hermetically sealed parts—that is, abusively separated—as well as that which has been *welded together*—that is, excessively connected. Our approach does not reject the principles of ecology. We do not want to define—in the manner of the currently flourishing "object-oriented ontologies" that we will analyze later on[45]—objective essences immunolog-ically cut off from any relational network whatsoever; rather, we want to complicate the principle of principles and inject into it a *counterprinciple of sep-aration*: a second principle setting out to transform the all-too-imperial econ-omy of the first principle. Such is our ontological therapy.

But the question is not simply an ontological one; it is also an eminently po-litical one. An ecology of separation affirms that, without some kind of *distanc-ing* within a socioeconomic situation, no real political decision is truly possible, no technological decision is truly foreseeable. When everything is considered to be continuous and connected without a rift or outside, automated reactions replace decisions, and each novelty that emerges into the market of saturated anthropogenic environments is presented as an inevitable fate. Incapable of thinking separation, eco-constructivism can do nothing more than evaluate, in a kind of pragmatic way, the effects of human actions within an "uncertain" world. "Uncertain" for those who don't strive to understand the inevitably contingent causes, arising from the decisions that could have been different than those that were taken and that have led to the installation of the world with which we are familiar. In this sense, in order to truly be political, to truly take into consideration the dangers threatening us, to distinguish between what humans can construct and what they can't and shouldn't construct, in order to know what is left for a possible usage of the word *nature* and the word *environment*, in order to create out of resilience a capacity for stepping back and regaining some distance, before simply being an aptitude for survival, ecology must leave a place for separation.

In other words, geo-constructivist extraterritoriality is a poor translation (a caricature) of the ontological separation necessary for any relation. In fact, there is only a relation where beings recognize their incompleteness, that they're lacking something, and as a consequence are called by the other toward the other; but this separation is transformed into a radical distancing establishing the position of a *rogue, quasi-off-planet* subject [sujet hors-bord] striving to vampirize natural dynamism in order to refabricate everything there is (geo-constructivism), or it is a separation that is refuted and that turns everything into "actants" inside "networks" of naturalized artifacts and of artificialized nature (eco-constructivism). We must indeed return to Earth, but without returning into the net of a flat world, rendering all beings equivalent and annulling all exteriority. In order for the return to Earth to not be of the order of a burial financed by the conquest of geo-capitalism, it is necessary to *metabolize separation*: neither a radical distancing nor an absolute interconnection, but an *interior distance* thanks to which an ecology of separation opens the Earth up to its own outsides.

FROM MULTINATURALISM TO ANTIPRODUCTION

It is in the name of this ecology of separation that we propose, during the third part of our book, another approach for nature. Certainly, it is necessary to refute the idea of some kind of "pure" nature. We can take the emblematic example of the *wilderness*, this so-called "wild nature" that would be present in the National Park system in the United States: Within quasi-sacred spaces, it is possible to undergo the experience of the sublime and re-establish some kind of contact with the pioneering spirit, the spirit of a time when the nature of the "new world" was still virgin territory. And yet, the *wilderness* is a "human creation" (William Cronon). It is a cultural construction.[46] In truth, the pioneers never traversed virgin territories: Natives had already inhabited those territories for at least ten thousand years. In 1492, between three and four million people inhabited the territory now constituted by Canada and the United States; but 90 percent of this native population was decimated as a result of the brutal treatment inflicted upon them by the Europeans, as well as a result of the diseases these Europeans brought with them. A complete reversal of perspective: In the nineteenth century, the portion of nature that appeared as virgin territory ripe for transforming into national parks was in fact the result of a complete and total erasure of operations conducted by the natives that had taken place over several centuries.[47] In a more general manner, we must get rid of the doctrinal and ideological usage of "nature" when it is used to justify racism, sexism, and fervent nationalism.[48] During the fight for marriage equality—*marriage pour tous*—that took place in France in 2013, Thierry Jaccaud, as the editor in chief of *L'Ecologiste*, declared the law put forth by the government as being "an attack against nature."[49] Despite these sorts of deplorable positions in the name of nature, this present book nevertheless maintains that it is important to go beyond the deconstruction of nature.

In fact, this deconstruction of nature is not merely theoretical but becomes socially and economically materialized within the incessant cycles of deconstruction and reconstruction that from now on define our planet. Remaking is the categorical imperative of our times, a disguised fantasy that has slowly become the program of an open sky, the backbreaking labor of a postmodern Sisyphus condemned to remaking yesterday what he should have modified tomorrow. Razmig Keucheyan is right in affirming that "nature is a battlefield," a "*theater* [my emphasis] of confrontations between actors of divergent interests": We must treat so-called environmental problems without dividing them up into issues of social class, gender, and race—such is the beginning for any environmental justice.[50] But the artificialist option that a number of Marxists share along with the constructivists—considering nature as a "theater," an ideological illusion, a mystifying abstraction, etc.—is precisely one of the forms of ontological ignorance that leads to occulting a part of the world. Removing the ideological veil from nature has become a new veil of an all-powerful constructivism that unremittingly deconstructs and reconstructs. At a time when the ecosphere is truly—and not simply conceptually—becoming degraded, it becomes literally vital to affirm that nature is not *merely* a battlefield: It is first and foremost a relational and intensive field of life and nonlife, which cannot be reduced to the measured and gadgetized spaces of human powers, inevitably escaping the grasp of strategic definitions proposed by the combatants, whether they be the definitions of the military or those of insurance companies transforming natural catastrophes and epidemics into financial markets.[51]

But what do we mean by what we just called "a relational and intensive field of life and nonlife"? What idea of nature are we proposing here?

1. An initial solution would consist of envisioning this field as a diversity of natures, a biodiversity in the proper sense of the term, which would include—among others—uncultivated or abandoned nature, nature that wouldn't exactly be pure but that would have once again become wild by way of the withdrawal of human activity (such as the area around Chernobyl), hybrid nature, domesticated nature, artificialized natures, etc. Largely in vogue these days, this thought of nature in terms of *structural diversity* could seem rather opportune: breaking up into pieces—literally—the idea of a single Nature, one that would be pure and substantial. This thought of a diversity of natures is welcoming, democratic, open to cultural singularities, including being open to cultures that do not seemingly recognize the so-called Western category of nature. But diversity always has a price: It also conjures up everything that would place a good diversity at risk. In making that which is wild an element among others, the *multinaturalism of diversity* cancels out its very scope, appearing with one hand to accept—diversity—that which it then refuses with the other—diversity, yes, but a diversity that has removed the real dimension of the wild. And yet, the category of the wild by its very essence spills over the boundaries of the place one wants to grant to it within the heart itself of diversity: Neither pure nature nor a wild, maddening weed, surviving and growing over the edges of asphalt, the wild is the *distinguishing power* as such, the matrix of the Other, the expression of what rumbles under diversity generating different natures.

2. Indeed, it seems preferable for us to think of nature in *genetic* terms. This is exactly what Eduardo Viveiros de Castro's *multinaturalism of differentiation* invites us to do: "a post-plural theory of multiplicities." "We could say that Amazonian multi-naturalism affirms not so much a variety of natures as the naturalness of variation—variation *as* nature."[52] And yet, for Viveiros de Castro, variation is unthinkable without a process of the actualization of the virtual or a becoming (in order to take back up the terms of Deleuze and Guattari used by the anthropologist), the passage from "an intensive superposition of heterogeneous states,"[53] into distinct worlds: the *incorporated world* of this jaguar, this plant, this human. In the way in which we render the work of Viveiros de Castro, Amazonian multinaturalism is the multiverse of distinct bodies that places on one side a common humanity from which beings are derived (in the beginning, as the Amerindian myths claim, all beings were human) and, on the other side, living differentiated entities (plants, animals, humans, spirits).

In this sense, *nature is the power of ontological distinction*. The objective of the ecological thinking that we are striving to promote is not, then, to deconstruct nature but to understand and accompany the genesis of the various forms of the natured.[54] We do not want to fill in the various great divides but rather change their meaning: Division, in its double meaning—that which distinguishes and which resembles—is *the common distinguishing power of worlds*.

This is what is truly at stake: thinking nature as a genesis and not as a structure. And yet, this book would like to contribute in thinking this genesis in another way than in terms of process, production, or transformation, as is all too often the case with how many thinkers employ the term. It's true, every genesis envelops an antigenetic layer, a phase of antiproduction, not firstly a *transformation* but a *withdrawal of form*. Instead of simply striving to know how we go from nature as (naturing) power to nature as (natured) object, instead of simply thinking nature as manifestation, brought to light, we will show that nature is a withdrawal, a contraction, a nonappearance that precedes the expression of beings and worlds. In contrast to the machinic ecology of Félix Guattari[55] and its geo-constructivist avatar, of an ecology of processes and hybrids, of an ecology that definitively reinforces the ontological management machine that underlies the will for managing the Earth-machine, an ecology of separation declares: Without an originary night, there would be no light of day; without a dark side, there would be no multiverse, and no nature in the plural.

It's this dark element of withdrawal that fundamentally escapes the grasp of humanity's power, the grasp of *Homo faber*. No doubt, humanity still attempts to hoard nature's forces, its naturing potential, humanity wants to make of itself a *Homo naturans*; but the naturing is always preceded by a denaturing element that is a lack of power or force [im-puissance] at the very core of natural forces [puissance]. This is the wild: not a privileged or banal space but rather a universal reserve of disconnection [déliaison] without which there would be no difference or diversity.[56] If every living being is exceeded by its wild character, including genetically modified beings that always tend to present unexpected behavior, if nature always separates itself from the way in which we try to nature it, this is precisely because the naturing is always exceeded by the denaturing, by the antiproduction preceding all production and that undoes, from within, the constructivist empire.

IΠ PRAISE OF THE UΠCOΠSTRUCTABLE

We should emphasize this point one more time: The constructivist empire is not merely a discourse, it doesn't simply concern a specific way of reflecting about "actors" and "networks" or a reflection about the way in which reality is "correlated" to human thought. The constructivism we're analyzing in this work is an *efficient constructivism*, truly producing the world and incorporating the discourses that accompany it or precede it in its attempt to integrally re-shape the planet. How do we contest this empire?

By localizing an unconstructable part that exceeds the flow of the development of the constructivist empire both upstream and downstream. *Construct* comes from the Latin *construere* (*cum-struere*): "to gather up layers," "to compile" in the sense of putting together (hence the use of the Latin *cum*, the *co-*, which signifies *with*). It is possible that the root *stru-* comes from °*ster-*: "to spread," hence the idea of "gathering up layers."[57] First and foremost, the unconstructable is what renders possible the act of edification, of building, or gathering together—a condition of possibility that in philosophy bears the name of the transcendental: not that which is transcendent, beyond, but that which, here down below, brings the world into being. And it is in this way that if we want to build something we must bring things together. But we can only bring entities together that were first of all considered as distinct unless building and edifying consists in bending a continuous plastic material without separation. To refuse this demented ontology, which is the plastic modeling clay ontology of capitalism, implies affirming that *the unconstructable is the transcendental of every construction*: that without which there would be no possible construction. This present book proposes that nature, in its denaturing form, is the transcendental allowing for the creation of a distance thanks to which it is possible to compose, construct, and to form. To think like a mountain, to refer to the famous saying by Aldo Leopold,[58] means getting away from oneself, to break away from one's immediate concerns, and in this sense, to withdraw: to contract to the point of creating a kind of outside the world inside the world thanks to which it becomes possible to understand the context of which we were stuck in the middle. Within a hyperconnected world where technology is used first and foremost to communicate, nature becomes not a means to "reconnect" with some kind of authentic and pure dimension of the universe; on the contrary, nature becomes a *vector of distance* capable of helping us to free ourselves from the viral character of electronic communication that envelops the world.[59]

By way of a conclusion to this present work, the category of the unconstructable will be used to rethink our relationship with the Earth. First, it is necessary to not get caught up in the official discourse of the Anthropocene, this new grand narrative that is presumed to provide some kind of meaning for the future fates of humanity. For there is no one single humanity: It is divided between those who have an interest in reformatting the planet and those who could suffer the disastrous consequences of climate manipulation—the entire problem of a truly political ecology is to highlight this division. In place of the modern myth of a conquering humanity that has become conscious of its power and that from now on is capable of devouring all of nature, we will help to bring into the picture *the double body of the Anthropocene*: One body of the Anthropocene is in a position of majority and is overtly technophilic, having established itself by way of an anaturalist position; the other body is composed of what we call *minoritarian bodies of the Anthropocene*—A body sacrificed on the altar of progress and its residue, a body that for several centuries now has contested the anaturalist position, a recalcitrant body for which the Earth is not simply some kind of hollow box but a full body, a body that knows scientific "controversies" first take the form of reckless experiments upon the territories of the living, a body of people overexposed to the industrial calamities in the Global South like those half-dead industrial areas in the Global North, a body privileging "democratic and diffuse" technologies centered around life against "centralized totalitarian technologies" (Lewis Mumford),[60] the minoritarian bodies of the Anthropocene form an alliance demanding climate justice.

In Ecuador and Bolivia, these minoritarian bodies highlight the importance of the "regenerative" capacities of "Mother Earth" and consider these capacities as a right: Throughout the world, they explain how and why these capacities of the Earth will always be out of the reach of projects of geoengineering; they make of the Earth a great animal—a living body, the organism of organisms, Gaia; what we will call a *full body*.[61] All of these neo-organicisms deserve attention and political support—however, are they enough for properly contesting the representation of the Earth as a *hollow box* favored by the geoengineers? In maintaining that the Earth is an "uncontrollable beast" that will react in unexpected ways to the projects of the geoengineers, are the neo-organicists doing nothing more than simply restating what the current Masters of the World keep repeating—namely, that the world is "turbulent," "uncertain," and that this situation is fated?

In order to properly respond to these questions, we want to propose a representation of the Earth that doesn't simply reduce it to an object (an empty box) or a subject (a full body); rather, we will think of the Earth as a *traject*. As a long-term event that began with the birth of our solar system, as an unconstructable trajectory, the experience of which cannot be recreated in a laboratory, the Earth exceeds the weak representation produced by the geoclimate technicians. But it also exceeds the holistic, organicist, and vitalist representation of the Earth as full body: Within the history of the Earth, the ecosphere is a very recent skin. No, the Earth is not Gaia, whether we consider Gaia to be a cybernetic system or a mother—a generous mother nature or an evil "intrusive," "dreaded," "quite indifferent"[62] mother, a terrifying monster who humiliates us.[63] Emerging from out of the obscure depths of our universe and destined for night, caught within an eccentric temporality displacing it from an objective or subjective spatiality, the Earth is an eccentric planet that temporarily aligns the nonliving with the living, very ancient forms and forms in becoming. The Earth escapes the grasp of humans in the way time passes and becomes concrete each time in an inimitable manner. Inaccessible from the beginning to the end of time, the Earth merely leaves a small amount of material for the geo-constructivists to decorate. Recognizing this unconstructable part of the Earth is what is required from a new political ecology: an ecology of separation for those beings who are neither astronauts nor beings frightened by the outsides of the world.

PART I

THE MIRROR OF THE ANTHROPOCENE

Geoengineering, Terraforming, and Earth Stewardship

We consider that which is possible as being absolutely obligatory, that which can be done as needing absolutely to be done. Today's moral imperatives are completely derived from technology.
—Günther Anders, *The Obsolescence of Man*, Vol. 2

Steven Austin, astronaut. A man who is barely alive. "Gentlemen, we can rebuild him, we have the technology."
—*The Six Million Dollar Man*

THE COPENHAGEN CHIASM

"This evening, the city of Copenhagen is a crime scene, with those responsible fleeing for the airport." It was in this manner that John Sauven, the executive director of Greenpeace for the United Kingdom, expressed himself following the Copenhagen summit on climate change.[1] A crime? What sort of crime? What exactly happened during this summit? More than likely, no kind of event that would be capable of immediately changing the history of the world. But nevertheless, there was a noticeable turning point in relation to how societies were discussing the management of climate change; there was a revelatory moment in regard to what we have taken to calling the *geo-constructivist program* and the globally dominant *anaturalist position*; the promotion of a self-enclosed technology, like a shield obliviously encompassing the world. As we will slowly begin to understand, the "face" of the Earth could be changed by this technology—literally.

But let's be less elliptical about it and simply present the facts. For the first part of this book, let's become *journalists of the spirit of the current times* and collect the empirical materials that will nourish our philosophical inquiry. For the countries present during this international summit, it was a question of negotiating an agreement before following through with the Kyoto Protocol. Throughout the discussions that took place during the conference, a so-called agreement was arrived at, recognizing, on a scientific basis, the vital requirement for maintaining the elevation of the Earth's temperature under two degrees Celsius, but at the same time placing no constraint whatsoever in terms of achieving this objective—that is, no formal measures for reducing the overall CO_2 emissions. One can understand why a number of newspapers—and not simply activists—spoke of the "failure" or "fiasco" of Copenhagen. But this initial evaluation—which was completely justified—nevertheless set the stage for a quiet "achievement" in the midst of becoming a reality: It's precisely during the course of this same climate change conference that the possibility of climate engineering began to carve out its own space of legitimacy.

While the question of geoengineering was not on the docket for discussion during the course of the summit, it could be heard being discussed everywhere in the background, in the shadows of the corridors, in the alternative presentations at the fringes of the summit, such as the series of panel discussions co-organized by the Royal Society titled "Science, research, and the international governance of geoengineering" and the projection of the film by Robert Greene, *Owning the Weather*.[2] The idea of a manageable geoclimate waited in the wings as a kind of ambush, sufficiently visible and audible to present itself as a spare solution just in case—in case the other possibilities ended up being impractical, in the event that the only thing the nations could agree on was the fact that it was impossible to come to an agreement on a concerted effort of drastically reducing CO_2 emissions. In the end, as far as the ability for nation-states to truly confront and discuss the true effects of climate change, the summit achieved a degree of intellectual altitude close to zero. This is how the former president Chavez analyzed the situation during his short speech in the plenary session: "If the climate was a bank, the wealthy countries would have already saved it," a comment that was echoed in vain by a number of associations that wanted to make the case for the necessary political constraints required in parallel with taking into account climate change: "Change the system, not the climate."[3] This exclamation is crucial and demonstrates the initial *chiasm* from which we can see the ideological-political position underlying climate engineering and, beyond that, all other projects with their sights set on reformatting the biosphere: Where activists for climate justice are pleading for politically changing the economic system leading to the techno-industrial modifications of the climate, climate engineering proposes technologically changing the climate so as to not change the political system already in place. Let's call this axiological inversion the *Copenhagen chiasm*; a temporal chiasm at a moment when a political possibility showed itself as being invalid. It is within this void that climate engineering was able to take on the character of a destiny.

CHAPTER 1

THE SCREEN OF GEOENGINEERING

A SHIELD OVER SUNRISE

Of course, this destiny was not forged in a day. The chiasm of Copenhagen is, as we have said, a fundamental turning point and not an absolute origin. Starting in 1955, von Neumann already began envisioning the possibility of a technological intervention in order to counteract climate change: "Carbon dioxide released back into the atmosphere through the industrial production of oil and coal has sufficiently changed the composition of the atmosphere to have caused global warming," he notes right after explaining that "the control of the weather and the climate," far from simply being reduced to a way for making it rain, will have as one of its principal stakes the control of the relation between the Earth and solar energy. Every climate phenomenon is

> ultimately controlled by the solar energy that falls on the earth. To modify the amount of solar energy is, of course, beyond human power. But what really matters is not the amount that hits the earth, but the fraction retained by the earth, since that reflected into space is no more useful than if it had never arrived.[1]

In 1965, one of the authors of a rapport warning president Lyndon B. Johnson of climate change proposed not to reduce the amount of CO_2 emissions responsible for this global warming but to modify the composition of clouds and to cast reflective particles out into the ocean.[2] In the scientific domain, the article that really sparked things off (if we dare to put it that way) is an article by Paul Crutzen titled "Albedo Enhancement by Stratospheric Sulfur Injections: A Contribution to Resolve a Policy Dilemma?" Very simply, the article details (following in a straight line from von Neumann's intuitions) increasing the reflective power (the aforementioned "albedo") of Earth by burning sulfur or sulfuric hydrogen injected into the stratosphere (thanks to cannons or balloons) in order to produce sulfur dioxide. These injections would help to counter global warming by way of *global dimming*:

> If sizeable reductions in greenhouse gas emissions will not happen and temperatures rise rapidly, then climatic engineering, such as presented here, is the only option available to rapidly reduce temperature rises and counteract other climatic effects.[3]

The failure of the Copenhagen summit will end up helping proposals such as solar shields to gain traction. Barely two days after the end of the summit, a certain Nathan Myhrvold—former technical director at Microsoft who now directs Intellectual Ventures, a company specializing in high-end technological inventions—was on CNN to speak about his "Stratoshield" that would use helium-filled balloons in order to suspend a tube vaporizing sulfur dioxide eighteen miles over our heads (in the Arctic).[4]

We will note in passing that Nathan Myhrvold is also vice president on the council of TerraPower, Bill Gates's start-up dedicated to nuclear energy—Bill Gates who describes geoengineering as simply an "insurance policy," something that he would prefer having in his back pocket in case "things happen faster."[5] It is true that capitalism can clearly hedge its bets on several different playing fields all at once: fossil fuels, nuclear energy, and—why not?—a cosmic hose. It's perhaps for this reason that however crazy the idea of Myhrvold's "hose-to-the-sky" might be, it was not used to make rain fall down on the fiery minds of those who, in the aftermath of the Copenhagen crisis, strived, by way of books and symposiums,[6] to already ideologically install a shield between us and the sun.

REAL SCREEN, IDEOLOGICAL SCREEN

An important question nevertheless remains to be answered: Why this specific technology? After all, geoengineering is a much more vast area of scientific exploration than the singular attempt of creating a solar shield with the help of a cosmic hose! The University of Oxford's geoengineering program defines this approach as "the deliberate large-scale intervention into the Earth's natural systems to counteract climate change."[7] For Oxford's program, geoengineering can be divided into the following two subcategories:

1. On one hand, there is *solar geoengineering* or what is also called *solar radiation management*. This technique has as its goal to reduce the quantity of solar energy that reaches the Earth, by way of a chemical shield or other kinds of techniques such as making clouds whiter (in order to increase their albedo) by

filling them with seawater or even painting large swaths of the Earth white so as to make it function like giant mirrors . . . The University of Oxford's program differentiates between "albedo enhancement" (by increasing the reflectivity of the clouds and the Earth's surfaces) and "stratospheric aerosols" (projections of particles in the stratosphere); but, as we have seen, someone like Paul Crutzen mixes these two subcategories together in speaking about "Albedo Enhancement by Stratospheric Sulfur Injections." We will conclude our discussion by mentioning the idea of "space mirrors," which, given their unlikelihood, do not even appear to spark the imaginations of Hollywood film studios.

2. On the other hand, there is *carbon geoengineering* or *carbon dioxide removal* that is also referred to by other groups as *carbon dioxide reduction*. This second category consists of any technique that contributes to the elimination of CO_2 in the atmosphere by capturing it and storing it—hence the synonymous term "carbon capture and storage"—in trees, which we should be massively planting everywhere, or through electric stations, or burying it underground or at the bottom of the oceans.

During the 2000s, it was the process of *carbon capture and storage* that attracted the most hope and a large financial backing. Already foreseeing that the nation-states would not arrive at some sort of agreement regarding the reduction of CO_2 emissions, instead simply wanting to maintain and continue development as it currently stood, the economico-political "decision makers," nation-states, and businesses, in the United States as much as in Australia, made the following wager: Let's continue to burn fossil fuels without any restrictions, but let's assume that technoscience will soon be capable of quickly removing the CO_2 from the atmosphere and be able to stash it somewhere. This wasn't what happened, and Clive Hamilton is correct in speaking about the first decade of the third millennium as being a "lost decade": Today, most of the carbon capture and storage projects have been abandoned.[8] It's true, the capturing of CO_2 would require an industrial apparatus that would produce an enormous amount of . . . CO_2! Our engineers have no doubt suffered from the belief according to which entropy would slam on its brakes at the gates of a modernity henceforth understood as "reflexive," capable of integrating into its calculations the dangerous waste products of its industry. However, it was still the right time to recognize the following material evidence: Given that the Earth is round and finite, all industrial activity on Earth will have industrial consequences . . . on Earth.

Goodbye carbon capture and storage. *Hello* climate engineering. If the method itself is different—preventing solar radiation from reaching CO_2 through constructing a chemical shield—the apparatus appears to be identical: continue to burn all fossil fuels (in other words, maintain the thermo-industrial capitalist economy at all costs). But climate engineering has a great advantage over carbon capture and storage: It allows for the maintenance of the capitalist economy without taking into consideration the anthropogenic consequences of climate change. Far from simply representing some kind of skepticism of the intellect, climate change denial still bears on the consequences that such a recognition would lead to: a complete challenge of the hegemonic discourse upon which the Anthropocene is constructed. And yet, in a certain way, climate engineering is the ad hoc production of a denial: On one hand, climate engineering recognizes that there is climate change; on the other hand, it shirks any human responsibility for it, since it proposes technology, industry, capitalism, and the possibility of being the master (and controller) of the Earth as the only solutions to our problems. Whereas carbon capture and storage had to, in spite of everything, confront the increase of CO_2 in the atmosphere by directly focusing its attention on the Sun, for which it wants to create a screen to counteract its effects, climate engineering turns its back on the Earth. A dangerous topology. With climate engineering, the real screen, the screen that reeks of the sulfur of technology, becomes superimposed with the ideological screen whose objective is to veil what we are doing. Such is, in the end, the meaning of this technology: to divert us, in the proper and figurative sense of the term, from a mode of civilization. "Reflexive" modernity sure does have a strange way of *reflecting* on its problems.

RHETORIC OF PLAN B

The perspective offered by climate engineering obviously elicits some strong objections, including on the part of the scientists promoting it. This strange reticence is one of its singular characteristics we must thoroughly examine. Take the case of Paul Crutzen and his well-known article from 2006, "Albedo Enhancement by Stratospheric Sulfur Injections." Crutzen was very careful to specify that it would be better to reduce CO_2 greenhouse gas emissions rather than simply release sulfur into the atmosphere, but for Crutzen, this reduction seemed to be nothing more or less than a mere "pious wish"—in other words, something that was completely unachievable.[9] Conclusion: Prepare for massive sulfur injections into the sky.

We could call this the *rhetoric of plan B*. This rhetoric presents itself under the guise of the best intentions in the world: Confronted with the dangers of rapid climate change perhaps being imminent, or in any case highly probable based on the feedback that has been recorded and other *tipping points* that will cause such irreversible changes in the climate because nation-states have revealed themselves to be incapable of fundamentally limiting CO_2 emissions, it will therefore be necessary to envision another solution—but which solution?

In 2013, Martin Rees, celebrated cosmologist from Cambridge, put forth the necessity for considering a geoclimatic "plan B" in case we were not able to greatly diminish CO_2 emissions within the next twenty years. In the eventuality of an extreme global warming of the entire planet, general overall panic would lead to emergency actions to be taken. A "plan B" accepting the "unavoidability" of our "dependence on fossil fuels, but fighting the negative effects of such dependence by way of some form of geoengineering" allows us to buy some time in order to develop more proper energy sources. And yet:

> Geoengineering would be an utter political nightmare: not all nations would want to adjust the thermostat the same way. . . . There could be unintended side-effects. Regional weather patterns may change. Moreover, the warming would return with a vengeance if the countermeasures were ever discontinued; and other consequences of rising CO_2—especially the deleterious effects of ocean acidification—would be unchecked.[10]

Rees's position is an eloquent one; eloquent in regard to the fundamental ambiguity of this so-called plan B. Since, generally speaking, we reserve a plan B *in case* plan A doesn't work out. As we have seen, Bill Gates compares climate engineering as a *just in case* kind of solution, just in case it would be necessary to create an immediate survival plan, if not for humanity, at least for Microsoft. The problem, however—hence our use of the term *rhetoric* here—is that *the promoters of plan B seem to not really believe in plan A*; in contrast, and this is the horrible paradox of this rhetoric, they know perfectly well the dangers residing in plan B, to which Martin Rees's comments attest. For the organizers of the conference Fighting Fire with Fire: Climate Modification and Ethics in the Anthropocene, which took place at the University of New South Wales (Australia) in July 2014, it is clear that the technologies used for intentionally modifying the climate are dangerous, and this danger has been exposed as such: "We're preparing to fight fire with fire."[11]

Indeed, at the very least, we can say that climate engineering is not entirely safe:

1. First of all, to the extent that the climate is a *system* that can only be understood as a set of interdependent elements, no *local* test can indicate what could happen *globally*; it would therefore be necessary to move directly to the stage of a full-scale, planetary test in order to have any possible idea of the potential consequences of a geoclimatic intervention—but are we prepared to become guinea pigs for such an experiment that would deeply modify our fragile and precious environment [milieu vital]?

2. Secondly, the atmosphere constitutes what physicists and mathematicians call a chaotic system: Its evolution is extremely sensitive to initial conditions,

and tiny variations in these conditions could provoke large disparities in regard to the final possible states of the system.[12] Moreover, it's for this reason that meteorological forecasting is so difficult to determine beyond two weeks in advance.[13] So we can understand the extreme difficulty of a large-scale test that would no doubt last several years: Assuming that a solar geoengineering project is agreed upon, the possibilities of an out-of-control technologically enhanced atmospheric system would always be present, as a shield that would risk falling on our heads.

3. Third, as Martin Rees remarks in the article cited above, it would be impossible to interrupt the activated geoclimatic program under penalty of seeing the climatic changes escalating beyond the rhythm they would have followed without a technological intervention.

4. More specifically, a number of studies have shown that solar geoengineering would affect the sky in a heterogeneous way. In other words, a modification judged to be positive in one location of the sky would be paid for at the cost of drastic changes in the climate elsewhere. The climatologist Alan Robock was able to show in 2008 that injections of sulfur dioxide would disrupt seasonal monsoons in Asia and Africa—they would reduce the necessary precipitation for supplying water to millions of people—and other studies show that several African countries would risk seeing a complete collapse of their crops.[14] So here we can see how geoengineering is connected to questions of environmental justice: This technology is clearly at the service of the Global North and those who want to participate in the geo-constructivist program without suffering the consequences. We have to remember when we're speaking about the Anthropocene that climate engineering considers itself as ready to save the planet—even if we have to pay for it by way of some collateral damage, such as with periods of severe drought in equatorial Africa and certain parts of India. Yes, the planet is a battlefield, but these battles are founded on the denial of the planet as a common entity *existing prior to any project of renovation.*

A FIREFIGHTER TECHNOLOGY?

As a first attempt at a synthesis, let's try to define what genre of technology has resulted from climate engineering, specifically in the form of solar geoengineering. Clive Hamilton qualifies it as a "technofix": a solution of technological improvisation that in no way transforms the causes of climate change.[15] As James Rodger Fleming writes in his book, *Fixing the Sky*, a "fix" is

> a dose of narcotics for an addict; or an illegal bribe or elicit arrangement. A fix is a measure undertaken to resolve a problem, an easy remedy, some-

times known as a "quick fix," which connotes an expedient but temporary solution that fails to address underlying problems.[16]

In fact, Fleming continues, it's the physicist Alvin Weinberg who forged the term *technological fix* in 1966. Since then, this term has come to describe

> simplistic or stopgap remedies to complex problems, partial solutions that may generate more problems than they solve. Placing more faith in technology than in human nature, Weinberg offered engineering as an alternative to conservation or restraint.[17]

Let's be more precise: Climate engineering is presented as opposed to "conservation," so, on the side of innovation and technological progress; but in reality, this technology prevents societies from transforming themselves, making any political or economical reform that would attempt approaching the cause of climate change impossible. When all is said and done, climate engineering is a way of forcing us to adapt.[18] But what exactly are we adapting to?

A new world managed by the new "minds" of climate capitalism: the geo-constructivists and their Promethean conception of technology, despite the fact that this Promethean label is perhaps not the best description. In a certain way, we could consider solar geoengineering as a kind of *firefighter technology*. Instead of acting in a *pre-emptive* manner, ahead of time, and directly applying itself to specific causes, a firefighter technology acts *after the fact*, on the consequences—which can only delight, as we will see later on, the partisans of a "pragmatic," "constructivist," or "modern" ecology that declares itself as only focusing on the consequences of our actions. Myhrvold's stratospheric shield, a hose set up in the sky, is the perfect illusion of the *fetishism of consequences*. But where we can see how this arrogant technology joins with Prometheus and his love of fire is through the fact that this technology wouldn't simply protect us from the Sun; it would also protect the "dominant minority" (Lewis Mumford) who set the planet on fire and led it to bloodshed. Under a sky whitened by sulfur, animal species will continue to die off and the oceans will continue to acidify.

CHAPTER 2

THE MIRROR OF THE ANTHROPOCENE

HUMANKIND AGAINST THE EARTH: A NEW GRAND NARRATIVE

In the beginning, there was Paul Crutzen: the prophet of the Anthropocene. And the prophet spoke: "Humankind has become a geological force." And his disciples chanted: "Nature has become human nature, no longer is there anything wild on this artificial planet that we must learn how to manage for the well-being of all of humanity." This modicum of narration is nothing other than the *mythological bedrock of the geo-constructivist discourse*. The master signifier of this myth is the "Anthropocene," and the objective of this chapter is to analyze the way in which the myth of the Anthropocene confronts two entities: the first entity is humanity or Humankind, the lone and superpowerful subject; the second is the Earth, an object that is just as unifying as Humankind but that is clearly in a subaltern position.

A reader might be disconcerted by our use of the term *myth*. By using the term *myth*, do we mean that all the sciences that are at the foundation of the term *Anthropocene*—the Earth sciences—are mythological? That is to say, false, purely ideological and unfailingly relativistic? No. What needs to be understood is the way in which the Earth sciences participate in an elaboration of *a general representation* of Humankind and the Earth that largely exceeds any scientific framework. This representation that relies on both the results of science and the imaginary can bear the name "grand narrative" or "metanarrative." Both of these terms were first used by the philosopher Jean-François Lyotard in the 1970s, who proclaimed the end of "grand narratives"—that is, narrations that aim to legitimize the production of the social reality.[1] According to Lyotard, the common function of myths and grand narratives is the legitimation of institutions and social practices; but whereas myths strive to uncover their legitimacy within an *originary* act (the transgression of divine edict, the theft of fire, etc.), metanarratives establish their legitimacy within a *future* that must be brought into emergence—the promised emancipation of Enlightenment Reason, a Socialism that would finally be achieved on Earth, Progress for all—in other words, any notion with a capital letter capable of providing some sort of meaning and direction for human civilization. With the Anthropocene, our winded postmodernity seems to have acquired a new breath and a means for resuscitating a grand narrative that, as we shall see, simultaneously plunges us into the most distant past—in the manner of a myth—and throws itself into an air-conditioned future of a manageable and controllable Earth. An absolute grand narrative that, in some ways, is simultaneously archaic and hypermodern.

Our book is certainly not the first—far from it—to focus its attention on the Anthropocene, a viral concept that now has its own journal.[2] If this concept has caught on and become rather trendy, like some kind of ideological portmanteau, it's because it allows for the performance of a very precise operation within the field of ideas concerning the relation of humans to the global environment: On one hand, the concept of the Anthropocene recognizes that *something is happening to us*, something important concerning the entire ecosphere, an event that involves the survival of humanity; but on the other hand, the designation of the Anthropocene has as one of its functions to transform *what is happening to us* by way of what human beings are *causing to happen*. For in the end, within the word *Anthropocene*, there is the word *anthropos*—that is, us humans, the species who is a specialist in making chemical shields. After all, *what is happening to us*, and what we are undergoing, is it really anything more than *what we are doing* and bringing into reality? Yes, we must protect ourselves—but we are simultaneously the danger and the remedy, a *pharmakon*, to take back up the Greek term meaning both the poison and what can also cure us. The geo-constructivist option obviously recognizes the existence of a climate threat, but it tends to mythologize Humankind's power—in other words, the remedy. A remedy strong enough to render the Earth definitively silent.

By evaluating this new grand narrative that is in the midst of becoming hegemonic, by analyzing the relation it weaves between science, history, and the imaginary, we will be able to understand why and how the Anthropocene legitimizes geoengineering as well as a certain representation of humans and what they are entitled to do. But we must also strive to escape from this grand narrative that justifies these despicable Promethean technologies: We must conjure the emergence of the multiplicity of scenes of the anthropo-scene, this comic and tragic theater in which the future of humanity on Earth embeds itself. If the Earth could speak, it would no doubt propose another narrative—a *geocene*.

HOMO SAPIENS, THE FARMER, THE INDUSTRIALIST, THE MAN IN A HURRY, AND THE COLONIZER

As with any representation in the midst of becoming hegemonic, the Anthropocene is subjected to a harsh ideological battle whose stakes are the following: What will be the victorious signified that will have the honor of filling in for the empty signifier of the Anthropocene? The term *Anthropocene* was actually invented by the ecologist Eugene Stoermer, who wrote, along with Paul Crutzen (a specialist of atmospheric chemistry), the famous article from 2000 simply titled the "The 'Anthropocene.'"[3] For these two authors, the transformation of humanity into a "major geological force" (due to an anthropocentric slip of the tongue, we always forget that the adjective "major" nevertheless limits the extent of the "force" in question) would be a primary effect of the Industrial Revolution. In other words, the Anthropocene would have actually started at the end of the eighteenth century with "James Watt's invention of the steam engine in 1784," write Crutzen and Stoermer (even though it turns out that 1784 might not perhaps be the precise date).[4] It's true that the analyses of air trapped in polar ice show a sharp rise in the global concentrations of carbon dioxide and methane in the eighteenth century.

Nevertheless, there are four competing versions that propose another interpretation of *Anthropocene*. These versions are not in agreement on the response that needs to be made to the following question: At what moment should we claim the beginning of the great narration of Humankind as a geomorphological force, capable of giving form (*morphé* in Greek) to the Earth?

1. After all, one could make the argument that, since its emergence around two hundred thousand years ago, humanity began its own reformatting of the Earth once it began to set fires, hunt, and merrily decimate the megafauna (European mammoths, saber-toothed tigers, etc.). This version of the Anthropocene, that incriminates *Homo sapiens* as such, would bring a tremendous amount of joy to certain *deep ecologists* who consider humanity as a cancer or plague [vérole][5] of which the Earth would be better off ridding itself . . . We could qualify this hypothesis as being hyperbolic, to the extent that it exaggerates the *initial* impact of humanity on the environment: Having an influence on the environment and transforming it is one thing; becoming a major geological force is another.

2. One could also posit the Anthropocene starting with the Neolithic period: this is what William Ruddiman, a paleoclimatologist from the University of Virginia, proposes with his "early anthropogenic hypothesis." This hypothesis, which clearly borrows from the work of Crutzen and Stoermer (but without naming them), posits agriculture as the origin of anthropic climate

change. It maintains that concentrations of CO_2 actually began to increase eight thousand years ago, with the practice of burning plots of ground along with deforestation in order to make way for a culture based around agricultural fields. As for the argument regarding the increase of methane levels in the atmosphere, the increase would have begun five thousand years ago with rice field irrigation and the overall increase in the raising of cattle.[6] Ruddiman in no way denies the vast changes to the climate resulting from the Industrial Revolution, but he asks us to consider anthropogenic climate change already underway during the Neolithic period—a warming powerful enough to perhaps have prevented a new ice age.[7]

The problem with this approach could perhaps be the same as the one preceding it: Can we compare an influence, however considerable it may be, with the way in which humans (as Ruddiman himself states several times) took control over the climate? However, as Christophe Bonneuil and Jean-Baptiste Fressoz claim, do these other approaches not forget to take into account the considerable large-scale changes marking a rupture between the period before and after the Industrial Revolution?[8] Does not erasing these changes merely affirm that the Anthropocene began when we became farmers?[9] The second problem, which flows out of the first, is the way in which Ruddiman situates his own discourse: on one hand at a distance from "environmental extremists" who—my word, how could they think such a thing?—"claim that the hands of industry spokespersons are soiled by financial support from coal utilities and oil companies driven by greed,"[10] and on the other hand, "extremists" of the industry who, "in response to alarmist statements" of environmentalists denounced earlier, "portray the Earth as resistant to the puny impacts of humans."[11] In caricaturizing "advocates of the Environment" as those who supposedly model their philosophical position on that of Jean-Jacques Rousseau and his conception of the "noble savage" and the idea of an "immaculate nature" worth protecting,[12] and describing those working in the oil and coal industries who, in the name of a "counter-offensive," see themselves as obligated to maintain an optimistic outlook, Ruddiman doesn't really seem to keep the proper distance from either group.

3. Or inversely, we could take up the position that the Anthropocene began much later than Crutzen and Stoermer claim: not two hundred thousand years ago or eight thousand years ago, but after World War II, by way of the sudden anthropogenic impact on the Earth. For example, it's true that there was a sudden increase in the amount of synthetically produced nitrogen fertilizer in the 1950s; and we could even begin to speak of a vast increase in the global population, water use, modes of transportation, mass planetary-wide tourism, and global telecommunications starting as well in the 1950s.[13] Global civilization

in its current form is the result of this coupling with an intensified agriculture in order to respond in a mega-industrial way to the increase in global population.[14] Nevertheless, we should not confuse an acceleration with an origin. Not that we have to contest the idea of this acceleration. And Crutzen, moreover, is completely ready to use the expression "great acceleration" if this expression describes the global situation shaped by new technologies, the limitless exploitation of fossil fuel energy, and the rapid increase in population.[15] But here we are confronted with exactly the opposite risk than that induced by Ruddiman's thesis: not a historical *flattening* linked to not taking into consideration changes at scale, but a shrinking leading to the development of a restrictive vision of history, conflating a specific moment with a radical rupture.

4. In complete contrast to such a shrinking, and by way of identifying a true radical rupture, we will mention the thesis developed by Simon L. Lewis and Mark A. Maslin, who propose the paradoxical date of 1610—paradoxical by the fact that it marks a significant reduction in the rate of CO_2 emissions in the atmosphere, due to the decline in human population provoked by the arrival of Europeans into the Americas. The diseases transmitted by the Europeans, as well as war, famine, and the implementation of slavery, led the population to go from fifty-four (or sixty-two) million individuals to six million around 1650. This reduction in the overall population also led to a reduction in agriculture and the spread of fire, and hence there was an increase in the regeneration of forests, prairies, and wooded savannahs, and, as a result, an increase in the capacity for carbon absorption by vegetation and soil:

> We suggest naming the dip in atmospheric CO_2 the "Orbis spike" and the suite of changes marking 1610 as the beginning of the Anthropocene the "Orbis hypothesis," from the Latin for world, because post-1492 humans on the two hemispheres were connected, trade became global, and some prominent social scientists refer to this time as the beginning of the modern "world-system."[16]

Homo sapiens, Ruddiman's farmer, James Watt wearing the suit of an industrialist, the man in a hurry, and the colonizer: These are the five remarkable figures that break away from the great fresco of the Anthropocene, each of them shaped by a different narrative claiming the status as an explicative hypothesis of the current climate situation. We will, however, indicate an important difference between the first four narratives and the last one: 1) *Homo sapiens*, the farmer, the industrialist, and the man in a hurry all unfold a kind of slow-motion cinematic Hollywood conception of man—this being called forth to become an absolute geomorphologist, first influencing the environment, then modifying it to the point of becoming—by way of an acceleration—the master of the climate; 2) this mythic grand narrative is nevertheless burdened by the negative history told by Lewis and Maslin. If the historical epicenter of the Anthropocene is the emergence of industrial capitalism, whereby Humankind is the embodiment of the master of technologically assisted Time, this origin is preceded by a kind of somber precursor, a collapse that should invalidate any kind of euphoric discourse in relation to the Anthropocene. The genocide of the natives already announced the real ultimate end of the Anthropocene: its propensity toward irreversible damage to the conditions for life. In reality, the great fresco of the Anthropocene depicts a climatic capitalism whose primary activity is valuable as a metaphor for the very conditions of its possibility: empty the Earth of all its inhabitants.

THE EARTH IS AN EMPTY BOX

If the Anthropocene is a great fresco, the Earth is its easel. In the grand narrative of the Anthropocene, a narrative validated as much by geoengineers as the financiers who believe that the manufacturing of the planet is a new market with the Earth depicted as a hollow machine. A container without any contents other than the human beings who, thanks to the might of their geomorphological strength, can fill it back in according to their whims and desires. This is precisely what is revealed (in a symptomatic way) by a video, titled—as it should be—"Welcome to the Anthropocene," that opened the United Nations Conference on Sustainable Development in 2012: In the film, we see the Earth depicted as completely hollow, a purely digital envelope subjected to quantification, slowly but surely becoming filled in to become a "natural" Earth.[17] How should we explain this desire and representation of the Earth?

Crutzen and Stoermer consider the Industrial Revolution as the (historical) origin of the Anthropocene, but we must add that its (metaphysical) foundation is none other than the modern science of the seventeenth century that, with Galileo, Bacon, and Descartes, assures the mathematical possibility of the mastery and possession of nature and, therefore, the mastery of terrestrial nature as well.[18] This is exactly what the young Marx, a true prototheorist of the Anthropocene, understood perfectly: Swept up by a conquering science, industry makes of the human being a "species-being" capable of "producing"—of regenerating—nature in such a way that nature "appears as *his* work and his reality."[19] From then on, Marx writes in an incredibly prophetic manner, the human being, like a well-equipped god, will be able to "contemplate himself in a world he himself has created."[20] Yes, the Anthropocene is a grand narrative, a veritable anthropo-*scene*—that is to say, an imaginary support structure—a scene of an industrial theater, a gigantic mirror apparatus [dispositif] where the unity of the subject called "humanity" constructs and contemplates itself within the unity of the Earth object. And in this sense, climate engineering is a significant step within the process of "contemplation" described by Marx: The human species constructs a shield capable of preventing the Sun from disturbing its narcissistic contemplation.

In researching the genealogy of such a representation of the Earth, we will rightfully find ourselves discovering the well-known images of the Earth seen from space: the first image being *Earthrise* by *Lunar Orbiter 1* from 1966 (the first space vessel sent up into space by the United States to orbit around the moon), then there was the *Earthrise* taken by Apollo 8 in 1968, and the most famous photograph, *The Blue Marble*, taken by Apollo 17 in 1972. As the historian Sebastian Vincent Grevsmühl writes,

> The space age gave birth to a vision of the Earth as an artifact, a radically closed and limited object, therefore something manageable and modifiable.[21]

These images do not represent the Earth viewed from the great sky above, from the terrestrial atmosphere; rather, they represent the Earth from a vantage point where there is no longer a position of above or below. We can then understand why, as Benjamin Lazier reminds us, Heidegger was "dismayed" by such images that materialized what the philosopher conceptualized by way of the term *uprooting (Heimatlosigeit)*, the impossibility of inhabiting, of dwelling in this world.[22] For Heidegger, the image of the Earth viewed from space signifies the loss of any sense of one's abode, or home, of any definitive existential direction (which is what the term *Heim* means)—"This is no longer the earth on which man lives," he complains.[23] Long before the somber ruminations made by Heidegger concerning uprooting, Nietzsche had already questioned—not without anguish—the "death of God."

> Who gave us the sponge to wipe away the entire horizon? What were we doing when we unchained the earth from its sun? Whither is it moving now? Whither are we moving away? Away from all the suns. Are we not plunging continually? Backward, sideward, forward, in all directions? Is there still any up or down? Are we not straying as through an infinite nothing?[24]

In Nietzsche's text, this narrative is placed in the mouth of a "madman" who continues thusly:

> Is not the greatness of this deed too great for us? Must we ourselves not become gods simply to appear worthy of it? . . . —Here, the madman fell silent and looked again at his listeners; and they too fell silent and stared

at him in astonishment. At last, he threw his lantern to the ground, and it broke into pieces and went out. "I have come too early," he said then; "my time is not yet. This tremendous event is still on its way, still wandering; it has not yet reached the ears of men."[25]

We could claim that the geoengineers and in general all the geo-constructivists, whether or not they have read Nietzsche's *The Gay Science*, have—perhaps unconsciously . . .—understood something of this "tremendous event" that the philosopher speaks of: Are they not striving to become these gods who alone would be "worthy" of the action consisting of having killed God?! Indeed:

1. The representation of the world that the Anthropocene proposes, and that nourishes geo-constructivism, fully takes up the act of nihilism: As with any modern science, for there is neither an up or down nor any kind of hierarchy of values, each form of existence is an object among other objects. Human beings can impose their rules upon all objects. The Earth as well must be considered as an object—a stellar object certainly, but an object nonetheless. Richard Buckminster Fuller baptized this special Earth object with the name "spaceship"—in other words, "a mechanical vehicle."[26]

2. Geoengineering and its *technofix* are presented as the medicine capable of healing us of this nihilism. For the geo-constructivists, the planet Earth, as with all planets, is—as its etymology indicates—an errant star. The Latin word *planeta*, which means mobile star, comes from the Greek *planetes*, as in the expression *planetes asteres*, which denotes the movement of the stars in contrast to the apparent fixed stars. The word is derived from *planasthai*, which means to "wander here and there, to wander off the beaten path," in the figurative sense of "being uncertain, floating."[27] But the Earth, as Fuller reminds us, is a spaceship, so we can steer it! We can make the wandering stop! The geoengineers will be able to cure us of this Nietzschean angst, this anxiety born out of the Galilean revolution whose principal cosmological and geographical effect was the decentering of the Earth and the Earth's inscription into a homogenous universe without any hierarchy or special center or fixed point.[28] The death of the Earth preceded the death of God—which Nietzsche had understood when he emphasized "the nihilistic consequences of contemporary natural science": "Since Copernicus, Man has been rolling from the center toward X."[29]

3. If the Earth is an engine, then like all engines, it can be repaired; one can remove and modify some of its parts if it starts to become defective. Hence the title of Fuller's book, *Operating Manual for Spaceship Earth*. In the chapter titled "Spaceship Earth," Fuller explains the extent to which the *design* of the Earth—its position in Space, the presence of the Van Allen belts that filter the sun's radiation, etc.—is perfect. The one thing that is truly empty for Fuller is,

without a doubt, our own knowledge of the Earth: "the designed omission of the instruction book on how to operate and maintain Spaceship Earth."[30] Or, to put it another way, what is truly lacking at the control panel of our planetary spaceship is precisely the proper knowledge of the Earth itself. But this lack of knowledge is totally fine for Fuller, who explains that this cognitive lack has forced humans to learn how to manage and pilot Spaceship Earth. This method of piloting or managing the Earth brings Fuller closer to the general systems theory biologist Ludwig von Bertalanffy, whom Fuller references, as well as cybernetics—the term *cybernetics*, as its inventor, Norbert Wiener, reminds us, comes from the Greek term *kubernetes*, or "steersman," the same Greek word from which we have derived our term *governor*.[31]

To summarize: There is no doubt, as Fuller emphatically claims, "we are all astronauts"; thanks to this position of extraterritoriality, we can control the Earth—in other words, we can modify our environment according to our needs and desires. Yes, we have lost our *Heimat* (our dwelling, our sense of home); but Heidegger was wrong in drawing some sort of melancholic perspective from this fact: We can take pleasure from our so-called *off-planet* location and—oh joy!—we can rebuild the planet according to our desires. We are astronauts that have the fortune of contemplating ourselves from inside a planet that it is possible to transform at will.

THE PARADOX OF *ANTHROPOS* (SPECIES AND KIND)

Let's take a bit of a closer look at these astronauts: Who exactly are they? These people we have taken to calling geo-constructivists. But this is not at all what they call themselves. Actually, they refer to themselves as Humankind, humanity, or even *anthropos*. A metonymy that begs to be examined a bit more closely, because it is at the heart of the very definition of the Anthropocene, a term that is invisible by its very hypervisibility: The Anthropocene is the age of Humankind. It's Humankind that has become a "major" geological force— Humankind, which Humankind?

If we follow the analyses of the anthropologist Tim Ingold, we can claim that there are two ways to define human beings: as a *species* of animal or as a *condition* opposed to the animal.[32] The first approach is rather inclusive as it tends to analyze humanity from a biological point of view, considering the human being as a specific animal, sharing certain common traits with other living beings and distinguishing itself from them by way of other characteristics. This approach obviously follows the logic of the Darwinian revolution: Humanity has no essence, it is a "continuous field of variations" considering each living individual a unique being.[33] The second approach proves to be exclusive, tending to define human beings as exceptional beings (and not unique), according to a dualistic mode of thought that Ingold connects to a humanism opposing humanity and animality, culture and nature, mind and body, emotion and reason, instinct and art, etc.[34] This fundamentally contradictory dual approach is at the heart of what we named, in the introduction of this present book, the paradox of the Anthropocene, which from now on is perhaps better understood by simply referring to it as the paradox of *anthropos*.

It's true, within the grand narrative of the Anthropocene, humanity is thought simultaneously as a superpowerful subject and as a powerless victim. The superpowerful subject is a way by which to think humanity from the viewpoint of its *condition*, which implies a *divide* between humanity as kind—"Marx's generic being," Teilhard de Chardin's "planetized humanity," etc.—and the rest of the world—the ecosphere, the various environments, the animals, etc.—over which Humankind's power is exerted, Humankind's power to *cause* climate change. By contrast, humanity, each time it considers itself as a *species*, must take into consideration the collateral damage of the Anthropocene, the impoverishment of biodiversity, climate change, etc. Within the dominant version of the grand narrative of the Anthropocene, humanity as a species is always rejected for the benefit of humanity as a kind. The geo-constructivist discourse can pretty much be summarized in the following description: 1) humanity as a kind—it doesn't matter whether it was eight thousand years ago or sixty thousand years ago—without knowing it, changed the ecosphere; 2) this produced a number of unfortunate consequences for humanity as a species, as a passive object and not as an active subject; 3) now that we know that we are in the midst of the Anthropocene, humanity as a geo-constructivist kind can consciously decide to transform the Earth, repairing all the negative aspects of the present situation that were the effects of modernity's appetite for conquest, and thereby affirm our hypermodernity. It's by way of this dialectic, consisting of rejecting humanity as species, that geo-constructivism attempts to liquidate the paradox of *anthropos*.

In complete contrast to the geo-constructivist discourse, Dipesh Chakrabarty, in a well-known article on climate change, arrives at the exact opposite conclusion, claiming that the Anthropocene should lead us to consider ourselves as a species.[35] According to this historian, who was primarily known for his contributions to subaltern studies and postcolonial studies, from now on it is no longer possible to distinguish natural history from human history: The two histories blend together inside the anthropogenic forge. From this point forward, we realize that what we have called "humanity as condition"— that is, as a cultural being in opposition to other natural beings—no longer has any meaning here. Humanity is nothing more than a species, maintains

Chakrabarty. A species, nothing more than a species? Does this sort of reasoning not lead us stumbling down a path toward the most dangerous of biologisms? Nevertheless,

1) Chakrabarty reminds us that the Darwinian revolution promised an approach toward species in nonessentialist terms. Darwin's theory actually takes as its target the Aristotelian concept of, horizontal and static, *specific difference*, with its sights set on substituting for Aristotle's concept of the *depth of field* what Darwin called "the community of descendants," the movement of immanent variations in which *all* living beings are swept up.[36] Why, then, should we deny the fact that human beings are just as much invested by this depth of the field of the living?

2) Especially since this depth is, literally, accordingly to Chakrabarty's analysis, bottomless. Because the human species produced by the Anthropocene is a *negative* subject, "a figure of the universal that escapes our capacity to experience the world," the figure of a "negative universal history."[37] Spanning a past of at least several centuries (or perhaps even several thousand years), as well as a future of several hundreds of thousands of years, the scale of the Anthropocene exceeds our ability to have an adequate experience—a representation—of such a temporal expanse: We can have an experience of the *effects* of our actions but not of what would be a species-subject as a cause of these actions.[38]

When all is said and done, it's not biologism that is problematic within Chakrabarty's account, but rather it's still the same old question of the subject-One, as the *anthropos* of the Anthropocene. Furthermore, this negative subject does not seem to promise becoming a political subject capable of overcoming its status as a global climate victim. Nevertheless, let's retain what is useful in Chakrabarty's analysis: In giving back to the *species* its proper place as signifier for humanity, his analyses help to prevent the premature liquidation of the paradox of *anthropos*. What is the real geo-constructivist fantasy? It's not simply that Humankind stands and confronts the Earth head-on but that the Earth disappears or dissolves into Humankind. If by chance, humanity, as a kind, ends up becoming some sort of postnatural *anthropos*, rid of its meat-flesh, which, as a species, suffers ecological aggressions, then humanity will finally be done with the paradox of *anthropos*! This paradox would melt away like the ice sheets in the Arctic. Actually, as long as the impoverishment of biodiversity will not have cruelly devastated the planet, humans will continue to *grant themselves an ungraspable kind*: Humanity is only a species when it discovers—after the fact, after every ecological disaster—that it lacks power.

CHAPTER 3

TERRAFORMING

Reconstructing the Earth, Recreating Life

RECONSTRUCTING THE EARTH

Power is at the core of the geo-constructivist conception of the world. The power of humanity as a kind to reduce its status of a species to its smallest part; a power that takes place on a devitalized planet. But power can be understood in two different ways: in political terms, as a form of governance, a kind of management, as well as in its rapport with technology. But before examining the eventual conditions for a governance of the Earth, we must insist on the following point: The political dimension of geo-constructivist power cannot be understood without taking into consideration its technological dimension. Before being a biopolitics, geo-constructivism is a technopolitics. This technopolitics invents an *off-planet* position for humanity from which, as we have already noted, humankind would be able to rebuild the planet according to its own desires. In this chapter, we will descend into the engine room of the geo-constructivist grand narrative, inside the secret chamber of its most persistent and ferocious fantasy: recreating the Earth, reconstructing life on Earth—to the point of repudiating death. The genealogy of this fantasy traverses the imaginary of the Space Age, as if, secretly, the Anthropocene had somehow inherited this imaginary. Let's begin by studying the transfer of the imaginary from the Space Age to the Age of Man.

Of course, it's not a coincidence that we find ourselves speaking of the imaginary and fantasy, as there is indeed something fantastical haunting the geo-constructivist soul, since reconstructing the Earth is derived from a term that originally comes from science fiction: *terraforming*. Jack Williamson was the first to coin the term in a short story called "Collision Orbit," published in 1942. In the story, the author imagines "space engineers" transforming an asteroid into an inhabitable place, thanks to an ingenious "para-gravity" system capable of rendering the atmosphere of the stellar body heavier.[1] Aside from this example, we can define *terraforming* as the operation consisting of rendering other stellar bodies—mainly planets and eventually asteroids— appropriate for human life. What is at stake is the possibility for another planet to become inhabitable, by way of modifying its ecosphere in such a way so as to make it possible for human beings to survive there, or in creating this life-sustaining ecosphere from scratch, so as to resemble the Earth.

Far from simply limiting itself to an imaginary arising out of science fiction, the idea of terraforming appears to have seduced many scientists for reasons that are intrinsically linked to the Anthropocene and to questions about geoengineering. In *Pale Blue Dot*, published in 1994, the famous astronomer Carl Sagan describes humans as vagabonds, nomads who left their birthplace of Africa and who would, one day or another, leave the cradle of Earth. For Sagan, it's not simply for ontological reasons that human beings will have to answer the call of leaving Earth for outer space but also for ecological reasons: "As the Earth's climate changes in the coming decades, there are likely to be far greater numbers of environmental refugees."[2] Therefore, we must take back up our baton as pilgrims and head off in search of other planets to which we can migrate. But Sagan adds that, in contrast to the migrations of peoples that take place on Earth, including those migrations to the New World of North America several hundred years ago, the migrations into outer space will face desert lands: "There are no distant relatives, no humans, and apparently no life waiting for us on those other worlds."[3] In other words, life will have to be created—or, at the very least, it will be necessary to create the conditions thanks to which life will be possible. Chapter 19 of Sagan's book, "Remaking the Planets," is therefore focused on terraforming. Sagan goes through a variety of possibilities including terraforming Venus (by way of recooling the surface of the planet, by creating, for example, a nuclear winter through injecting the pulverized debris of an asteroid into Venus's atmosphere), Mars (by way of creating a greenhouse effect that would be catastrophic on Earth but a godsend on Mars), or even the moons of Jupiter. It may appear as a troubling fact to read today that already in 1994, Sagan clearly saw the connection between the hypothesis of terraforming and what we now call the Anthropocene:

> We need look no further than our own world to see that humans are now able to alter environments in a profound way.[4]

Sagan clarifies that these changes—the depletion of the ozone layer, global warming, etc.—happened because of inattention, whereas now it's possible to produce these effects intentionally. In this sense, *terraforming would be noth-*

ing more than controlled anthropic terrestrial alteration. We should also note that Sagan is well aware of the fact that "some of the techniques that might eventually terraform other worlds might be applied to ameliorate the damage we have done to this one": Sagan, the great lover of otherworldliness, tells us that we must create a life-size "test" of terraforming on our world, guaranteeing its inhabitability, before using these techniques on other planets in the solar system.[5] Strange mirror reversals, where the terraforming of other planets and the anticipation of strictly earthbound geoengineering exchange their qualities and objectives: Is the future of humanity truly beyond Earth, or should we indeed rebuild our planetary cradle?

THE REVERSAL OF THE FRONTIER

Sagan writes that during the hunter-gatherer period, "the Frontier was everywhere," but we have abandoned nomadic life. From now on, we can find this form of life, again, in outer space. Sagan cites Melville's *Moby-Dick*: "I'm tortured by an everlasting itch for things remote. I love to sail forbidden seas."[6] Frontiers, galactic pioneers, distant lands—in his argument, often laced with hints of lyrical romanticism, Sagan in fact appeals to a myth that is deeply anchored within the historical imaginary of the United States of America: the "national myth of the Frontier," as it is also named by William Cronon.[7] This myth deals with the manner in which the great movement of the pioneers toward the West was represented during the nineteenth century: as if they had traversed and truly encountered a wild part of nature, never before touched by human beings. The historian Frederick Jackson Turner argues that the "end" of the Frontier—the end of expansion toward the West—had been replaced by America's psychopolitical investment in the *wilderness* as a new "wild" space: The *wilderness* became the avatar of the Frontier.[8] As the historian Howard E. McCurdy demonstrates in *Space and the American Imagination*, the arrival into the Space Age, symbolized by the inaugural launch of *Sputnik* in 1957, became a new incarnation of the Frontier myth. Space, McCurdy writes, was described, dreamed of, and constructed by the American government, the popular press, within the movies and TV shows, as well as scientific discourse, as the "final frontier," as the promise of new virgin territories to conquer, a horizon as distant and vast as that of all of North America and its appetite for Melville's "forbidden seas."[9] And with this final frontier, there will more than likely be no natives to eliminate—except, perhaps, some very rudimentary subterranean life-forms. Just a bit of anticipated terraforming to plan for that can quickly be put into place following the likes of the representations proposed by Robert McCall, known, among other things, for his illustrations of the Space Age—for

FIGURE 1. *The Cosmic Ark,* a colony suspended in the void thanks to artificial gravitation. © NASA Ames Research Center, Rick Guidice.

magazines such as *Life,* for NASA, and Stanley Kubrick. An alternative solution: create some sort of "cosmic ark," a colony that would be suspended in the void thanks to an artificial gravitational system (like the one we still seeing being proposed in the film *Interstellar* in 2014, see Figure 1).[10]

What sort of future did the promoters of the "Final Frontier" anticipate? In 1986, the report of the National Commission on Space, presided over by a former director of NASA, recommended the installation of an advanced post on the Moon by 2006 and humans on Mars by . . . 2015: "Many of the people who will live and work on that Mars Base have already been born,"[11] the authors of this report shrewdly claim.[12] Among the arguments put forth in order to garner excitement about extraterrestrial colonization would be the possibility of loosening the ecological constraints on an Earth that is becoming more and more populated, where the competition for limited resources was only going to grow.[13] But the future has been more complicated than we anticipated. The last time an American made it as far as the Moon, notes the journalist Elizabeth Kolbert, was in 1972![14] The Frontier lowered itself into the terrestrial orbit of the International Space Station launched in 1998. A number of ambitious projects have since been abandoned, such as the new United States space shuttle scheduled for completion in 2011, or the Constellation program that targeted a return to the Moon in 2020 followed by an inhabited flight to Mars.[15] Little by little, desire and the imaginary deserted the deserted space of outer space. But why?

Firstly, the technoscientific emulation generated by the Cold War—*Sputnik* vs. Apollo—tumbled to the ground along with the Berlin Wall. The end of the USSR, as the astronomer Chris Impey notes, led to a collapse in research funds for Soviet space exploration.[16] As for the Americans, a series of repetitive traumas—the *Challenger* crash in 1986, then the crash of the *Columbia* in 2003 (seven deaths in both cases)—have not helped the situation: The second death of Christopher Columbus (the spaceship *Columbia*) wasn't a good omen. Far from being a frontier leading toward the propagation of life, the Space Age transformed into a wall upon which astronauts could crash. The conjugated results from this historical period and its disappointments: a drastic reduction of NASA's budget. Hence the title of an article from the *Economist* in 2011, while the United States buried—if we can put it that way—the project for a new shuttle: "The End of the Space Age." The article specified that "Inner space is useful. Outer space is history."[17] The abandonment of the Space Age as a grand narrative led to what we will refer to as the *Reversal of the Frontier*. The psychopolitical investment of the conquest of space during the Space Age was transformed into an investment regarding the conquest of the Earth: a hypermodern conquest in search of an Earth 2.0 that is not outside of geostationary orbit but rather resides on the workshop table of geoengineers.[18] An Earth that has become neither virtual nor digital but an Earth that has become an *augmented reality*.[19] In a certain manner, this reversal or U-turn has consisted of representing the Earth as if it was another planet, like an exoplanet, as some other stellar body without any specific qualities, as some kind of material without any relation whatsoever to our own human history. In this sense, geoengineering, as it has been promoted by the geo-constructivists, is the inheritor of the projects of terraforming that have haunted as much the imaginary of science fiction as they have all of the postwar sciences. The end of the Space Age will have therefore been one of the historical conditions necessary for the birth of the Anthropocene: For this grand narrative, humanity is external to Earth not simply because humanity considers itself as some kind of nonliving entity but also because the *Earth is considered as being nonterrestrial*. Going from the Space Age to the Age of Humankind, the grand narrative of the Anthropocene took over the symbolic conquest of Space: reconstruct the climate, reconstruct everything in an anthropic way, and without leaving the Earth. A sort of domestic terraforming.

It's true that certain private companies have taken the baton passed to them from the nation-states, such as Virgin Galactic, who wants to "open space to everybody."[20] Led by the highly media-savvy Richard Branson, this company sports an apparently unrelenting faith in the conquest of space: "we aim to transform the current cost, safety, and environmental impact of space-launch. In doing so, we are helping to create, for the first time, a basic space access infrastructure that will act as an enabler for scientists and entrepreneurs. It will also provide a catalyst for a new age of space exploration which promises enormous positive potential for life on Earth."[21] As for Elon Musk, Silicon Valley's rising star and head of space company SpaceX—following the proper geo-constructivist logic—the conquest of space is simply about saving the planet. In order to do this, on one hand, one should develop a so-called "green energy" (on Earth)—in other words, electric cars (which is the raison d'être of Tesla Motors, directed by Musk); on the other hand, for the human race to survive, one also must "become a multi-planetary species," establishing a colony on Mars as soon as we possibly can. Not without a bit of humor, Musk has even exclaimed, "It'd be pretty cool to die on Mars, just not on impact."[22] While it's still easy to find people, such as the writer Stephen Petranek, who believe that Mars "will become the new frontier, the new hope, and the new destiny for millions of earthlings who will do almost anything to seize the opportunities waiting on the Red Planet,"[23] these private space companies have also experienced their setbacks—such as the suborbital crash of the craft SpaceShipTwo (constructed by Virgin Galactic) in October 2014 (with one death) and the explosions of a Falcon 9 (SpaceX) in June 2015 and September 2016.[24] What's more, it appears that transporting human life into space is a rather difficult task, the interstellar void is frankly not that favorable to humans, and the task of adapting to another planet, while humans are pretty much made for planet Earth, is anything but obvious. The dream of finding a solution beyond the Earth is to forget, as Elizabeth Kolbert rather eloquently puts it, that "wherever we go, we'll take ourselves with us."[25]

In other words, the Reversal of the Frontier, which is the psychoeconomical operation at the heart of the Anthropocene, does not imply a drought of extraplanetary desire. Not a month goes by that we don't discover a new exoplanet that could possibly hold the conditions for life and the welcoming migratory flight of humans from Earth. And aside from a "plan B," and the ongoing discussions regarding geoengineering the Earth's climate, there is still talk of a "plan C," which would be the exodus to another planet.[26] But geo-capitalism doesn't seem to lend much *credit* to such an eventuality. Is this because the idea and representation of a "cosmic ark" has something of a déjà vu about it, like a ghost from the 1950s? Is it because such a project would contradict the terrestrial limits of the grand narrative of the Anthropocene? As a symptom of this problem, let's consider the special issue that the popular French science magazine *Science et Vie* dedicated to "extra-solar paradises," to superhabitable planets—that is to say, "lands more habitable than our own." The magazine tells us that there would be more than five billion superhabitable planets! That's something to get any colonizer excited! Unfortunately, further along in the same article, two other inserts indicate that: 1) there are no means for propulsion possible to allow us to arrive to these exoplanets—it would take five thousand years to make it to the closest one—and it's hard to imagine an ark that would journey longer than it took *Homo sapiens* to develop; 2) these superplanets would be wonderful for fostering life—but not for human life—the gravity would be too great.[27] The Reversal of the Frontier leaves the care of rebuilding the Earth to the geo-constructivists, whereas a bit of imagination is left for the construction of space shuttles condemned to monotony: circling the Earth in perpetuity without ever leaving the upper atmosphere for outer space. Concerning the real exoplanets, the fact that they are beyond our reach does nothing but confirm the necessity of geo-constructivism.

RECREATING LIFE: CREATION, SYNTHESIS, RESURRECTION

Terraforming is intrinsically linked to the stakes of the living [du vivant]. First for ethical reasons: Terraforming or any kind of colonization would alter the exoplanet, preventing the scientist from analyzing the extraterrestrial environment as it is, untouched by humans, and putting at risk the life-forms that would already potentially exist on the planet—such as micro-organisms.[28] Beyond these somewhat premature ethical questions, it's important to understand that the idea of terraforming is linked with one of humanity's oldest dreams: recreating life—that is, to artificially master the processes for the generation of life, even if this reformatting of life bears with it some form of collateral damage, like the extinction of an exoplanetary form of life.

It's not by chance that James Lovelock, the inventor of the Gaia hypothesis—namely, the theory by which the Earth functions as a kind of superorganism—is also the same person who wrote a book about terraforming Mars: *The Greening of Mars*, published in 1984. And yet, the Gaia hypothesis arose out of the work Lovelock did for NASA regarding the possibility of detecting life on Mars. To sum up this earlier work: If there is life on Mars, this life will use the atmosphere, will change it, and will transform it for its own use. If the atmosphere on Mars appears stable, without any signs of disturbance, then there is a large probability that there is no life on Mars. If the atmosphere is in a state of "non-equilibrium" and resembles a "metastable" system, one can bet on the fact that living beings would have produced this dynamic disequilibrium. From this fact, Lovelock concludes, the dry, flat aspects of Mars, the relative chemical equilibrium of the Martian atmosphere, doesn't leave us with much reason to think there is a presence of life on Mars.[29] Almost twenty years after its publication, *The Greening of Mars* turns the argument in question on its head: How do we create a dynamic disequilibrium that would make Mars *in fine* habitable for human beings? By starting with a terraforming project through an intensive bombing of the Martian soil in order to create an artificial greenhouse effect, followed by the colonization of living organisms that could transform the atmosphere for their advantage.[30]

Of course, the desire to *produce* life through and within artificial conditions is nothing new—from Pygmalion to the facetiae of Dr. Frankenstein. But with what we have called the Reversal of the Frontier—an influx of psychic, political, and economical investments, re-emerging from a Space Age considered as obsolete and moving toward an age of Humanity (re)turning back toward Earth—it's as if the production of life has become the principal piece of a will toward a limitless terraforming: rebuilding life as the principal dimension for reconstructing the Earth. From the *greening* of Mars, we move on to the *greening* of the Earth, as if the latter, in a certain way, seemed to be *lacking life*, as if the Earth were *already dead*. This is what, according to us, indicates the growing importance granted to synthetic biology. A science undergoing a great expansion and unleashing in its wake a colossal financial investment,[31] synthetic biology was defined in 2010, by a commission in the United States, as "a scientific discipline that relies on chemically synthesized DNA, along with standardized and automatable processes, to address human needs by the creation of organisms with novel or enhanced characteristics or traits."[32] Synthetic biology could finally be used to conceive of new ways of creating GMOs, for producing new medicines, new biofuels, bacteria that could help us to diminish our use of toxic substances, even creating lightweight synthetic spider silk that could be used in the aviation or automobile industries.[33] Established on the idea that the living is reducible to "bricks"[34] (BioBricks), synthetic biology is highly compatible with the ideas of geo-constructivism: We could use it to address the problems of climate change as well as energy and food shortages thanks to an almost miraculous technological solution.

With its "enormous power for altering life on Earth,"[35] synthetic biology is certainly not without its potential dangers: New organisms, such as GMOs in general, will always be susceptible to unbridled proliferation, of mutating and replacing gray pollution (industrial pollution) with an "augmented" green pollution (a genetic pollution). More generally, synthetic biology rebels against, and is in radical opposition to, any kind of idea of *conservation*—of the environment or of a species: Why protect what we can improve, or reconstruct?[36] This sort of substitution of conservation by synthesis is particularly visible within current projects on "de-extinction" that would allow for the resurrection of certain species that have already recently disappeared (such as the passenger pigeon and the dodo bird) as well as species that went extinct long ago (such as the aurochs).[37] Sometimes these projects are connected with a desire to "re-wild" certain territories such as the Oostvaardersplassen nature reserve in the Netherlands, supposedly with the intention of restoring a Paleolithic type of nature, a "new nature" whose traits would nevertheless be to resuscitate a nature that was already *dead and buried*.[38] However, the de-extinction of species does imply a minimum amount of conservation, since it is only possible by way of using intentionally preserved DNA—or some kind of fossilized DNA—belonging to a disappeared species, a conservation that could implicate systems of cryo-conservation (where cells are frozen at very low temperatures). However, can we claim, along with Stanley Temple, a professor who specializes in conservation and wildlife, that the very act of "re-creating" disappeared species would be an "unprecedented *biological* (our emphasis) event"? What actually is the status of the living implied within the activity of reviving "dead species"?[39] What does it mean to give life back to a Pleistocene landscape? Would it really be creating life? Or, rather, would it be a form of producing nonlife, some kind of form of the living-dead? Stanley Temple is correct: Recreating dead species is "no longer science fiction."[40] But that is precisely the problem. From now on, the hypothesis that "extinction will no longer be forever," and that now there will be new forms of invasive species, of species "coming from the past,"[41] is not science fiction but fiction that has been made into science: the materialization of technologically assisted specters.

In this sense, de-extinction should be interpreted less as a promise of progress than as a threat whose fallout we cannot as yet measure. Not simply for the obvious reasons that the desire for treating the causes of environmental disasters and the loss of biodiversity risk being abandoned for the benefit and profit of the merchant desire of reformatting life, but also because death itself no longer has a place within this scientifico-discursive apparatus. Of course, de-extinction will not really resuscitate disappeared species: It will genetically create new ones, in contact with environments that are completely different than those with which the disappeared species had actually been confronted. We get a much better depiction of the future in *Jurassic World* (dir. Colin Trevorrow, 2015) and its alternative past inhabited by hybrid dinosaurs than by *Jurassic Park* (dir. Steven Spielberg, 1993) and its cloned dinosaurs. And yet, even if de-extinction is for the moment still a fantasy, it takes nothing away—it's completely the opposite—from its effectiveness within reality: to conquer death, to rip away the mortality that clings on to living beings and to be able to toss it into the garbage heap of obsolete anxieties and fears, is precisely the aspiration that underpins the genetic acrobatics of our current era.

Finally, we are confronted with something that should force us to reinterpret Foucault's hypothesis regarding biopolitics as a modern form of governmentality, in the manner in which it began developing starting in the seventeenth century, consisting of "making living" and "letting die"—instead of "making die" and "letting live" in the manner of more ancient sovereign societies.[42] The problem for those we will call the bio-constructivists is not first and foremost to improve the living but to repudiate death. We could say that there is nothing new about this repudiation and that numerous cultures have been founded on the denial of death. But with the advent of de-extinction and its regeneration of zombies, it will become less a question of "making live" than of *re-making live*. The death of nature will simply have been the necessary step toward the programming of its controlled resurrection.

THE MANAGEMENT OF A RECONSTRUCTED WORLD

From now on, we find ourselves at the border between technological or biotechnological power and political power. In this light, how does one define the political modality that seems to generate synthetic biology? Catherine and Raphaël Larrère think scientists who practice biology, considering themselves to be some kind of demiurges, are in reality mere "managers" who don't realize it. Synthetic biology, as well as cloning and the production of GMOs, are really nothing more than forms of "sophisticated bricolage" with random results that, in the end, rely on nature and its "possibles."[43] In fact, these practices are not a result of the art of "making" but the result of the art of "making do-with" (or "making to make" [faire-faire], which is merely a variant of the "making do-with"): What's at stake is not a "mastery" but a "steering or management" [pilotage]—that is to say, a capacity for "stimulating, governing" the capacities of the living.[44] Within the model of steering or managing, "we consider nature as a partner, we collaborate with it."[45] To definitively state that these practices are "explorations of the natural possibles" seems to put in check the exaggerated pretensions of the engineers for mastering life, but as a result, this idea risks leading to the exact opposite effect.

1. In the preceding sections, we wanted to demonstrate that technological mastery is not measured simply by way of its effectiveness within the achievements made by the priests of the laboratory but by way of what this mastery engages with in terms of material representations of the world and the effects of these representations. Mastery is, first of all, the mastery of the various

discourses that led to the privatization of the living from which future enterprises of the resurrection of the living can benefit. All of these activities are not of the order of the natural *possibles* but of the order of economic *necessities*. Once this is understood, to say that nature is a "partner" within the ontology that leads—or not—to the results of *scientific* efforts, is this not a way of definitively idealizing a socioeconomic process that is massively asymmetrical?

2. In a more profound manner, we wonder if the metaphor of the French word *pilotage*—meaning steering, managing, or piloting—is truly sufficient for contesting the position of the demiurge. It's true, steering or driving (piloting) a car is to steer a means of locomotion that is completely fabricated. In other words, as we will see, the bio-constructivists or geo-constructivists are in perfect agreement for taking up the task of steering or piloting (managing) the ecosphere in the same way that Buckminster Fuller, as we have shown, believed in our abilities to pilot "spaceship Earth." All in all, and following its etymology (borrowing from the Italian *piloto*, from the fifteenth century, or *pilota*, from the sixteenth century), "the one who steers a boat,"[46] steering concerns, first of all, that which is outside of nature—the machinic or the cybernetic as the lack of the distinction between the living and the artificial for the victorious benefit of the latter. As a metaphor, steering or managing implies a position after the fact—following the implementation of a *steering mechanism* [machine de pilotage]—that doesn't seem capable of keeping in check the increase in geopowers and their appetite for an integral reformation of the Earth.

CHAPTER 4

THE LOGIC OF GEOPOWER

Power, Management, and Earth Stewardship

Remaking the Earth, modifying organisms, resuscitating species. Each of these activities must be apprehended as a specific *economy of the world*: the management of environments and their increasing modification, with its sights set on resounding and startling benefits. In this sense, the Anthropocene—its geoclimatic repercussions and aspirations for a complete overhaul of terrestrial nature—denotes a vast worldwide techno-commercial operation whose aim is what climate specialist David Keith[1] calls "planetary management," which would, as Clive Hamilton remarks, allow for managing the "remainder of the natural world like a garden."[2] It's as if the numbing of the Earth—in the same way one numbs a tooth before pulling it—was one of the necessary conditions for allowing geoengineers to justify their mode of intervention. As the American astrophysicist Lowell Wood (a prominent figure in questions and research around geoengineering) puts it: "We've engineered every other environment we've lived in, why not the planet?"[3] There is no doubt, the Anthropocene is a self-validating discourse: 1) it posits *everything* as being human, that everything has been transformed by humanity as kind, as a "species-being," and as a consequence that we're in the age of Humankind; 2) so that it can authorize itself to complete the job: to humanize what would remain of the "natural world"—in other words, to *anthropoform* and manage the whole thing.

To complete the job, to humanize every last inch of the planet, to manage the whole thing: the term *stewardship*, a particularly fashionable term used often in current environmental research, helps us to understand these operations. According to the authors of the article "Earth Stewardship: Science for Action to Sustain the Human-Earth System," Earth stewardship is a "science that facilitates *actively shaping* [our emphasis] trajectories of social-ecological change to enhance ecosystem resilience and human well-being."[4] The fundamental idea that we will explore, throughout the second part of this book, is the ability to manage the Earth in its current and future states. The authors of this article tell us that it's preferable to anticipate a problem rather than have to remedy a problem. That is, a problem *for* humans: "The goal of earth stewardship is not to protect people from nature, but rather to protect nature for human well-being." So we will not be surprised to see nature considered as an ensemble of "services." As for geoengineering, the authors warn of its potential dangers: How can nations come to an agreement over several centuries in order to collectively manage the climate in spite of the political changes that these countries might experience? What would happen if a war broke out? Nevertheless, the authors consider that the "technological paradigm" is something not to be rejected but rather to be reconciled with the "paradigm of economic development" as well as the *adapting mosaic paradigm*, in regard to the resilience of ecosystems. The fact that the last of these paradigms carefully avoids using the word *ecological* in its definition is not mere coincidence: The stewardship of the Earth is supposed to define a perspective that is capable of integrating more metrics than those that have traditionally been taken into consideration by ecologists.

Are we trying to say, once and for all, that the idea of stewardship is, in itself, bad? Surely its ethical dimension, the demand for more responsibility that such an idea calls for, cannot actually be harmful! Nevertheless, we should recall that the word *steward* once denoted an agent with royal power (a financial steward, a steward of justice, etc.); in the eighteenth century, this name was applied to certain functionaries in charge of public organizations or services.[5] How has it come to pass today that we no longer seem to notice that the definition of Earth stewardship appears to have taken on the exact opposite of its original meaning! Whereas stewardship was supposed to be practiced for the benefit of *an other*, for *someone or something else* (the sovereign, the State, the public), with its geo-constructivist meaning, stewardship becomes *a practice for the benefit of the steward*—as if stewardship concealed within it a *self*-stewardship [auto-intendance]. In the end, the Reversal of the Frontier doesn't leave much room for alterities (whether extraplanetary or terrestrial) when these alterities aren't considered human in the first place. Because auto-stewardship no longer contains any sort of ethical dimension, it becomes indiscernible from some kind of *self-service* and, therefore, no longer allows us the convenience of being able to judge various technologies: It becomes difficult to resist the temptation of accepting climate engineering and its promises of an ecospherical thermostat; it becomes difficult to say no to something that presents itself as a possible improvement of our comfort. As a consequence, we won't be surprised to read articles that are clearly dedicated to researching a possible reconciliation between Earth stewardship and geoengineering, such as the article "Geo-engineering, Governance, and Social-Ecological Systems" by Victor Galaz, a political science professor and researcher at the Stockholm Resilience Center, who is working on questions of "global environmental governance."[6] This article is not that far from the "rhetoric of plan B" that we examined earlier: The author makes a list of all the horrors that could arise out of climate engineering, yet he still tries to demonstrate that "Earth stewardship and geo-engineering are not necessarily in conflict, but instead could be viewed as complementary approaches."[7] The final objective, announced in the last couple of paragraphs of the article, is crystal clear: "to shift from unintentional modifications of the Earth system" to "an approach where we intentionally try to modify the climate and associated biogeophysical systems to humanity's benefit."[8] Indeed, how can we resist the *power of attraction* of a geoengineering that has been generously devoted to humanity's good fortune?

The only way of resisting this temptation consists in using the concept of stewardship (as certain authors have done) by way of conserving within it a part of alterity: as not simply being a service for oneself. This is what the authors of the article "Global Assemblages, Resilience, and Earth Stewardship in the Anthropocene" tried to develop, for whom the concept of Earth stewardship is synonymous with societal obligations *to* nature.[9] The problem becomes the "decoupling" between these obligations to nature and those obligations that would only concern humans or socioeconomic systems. This decoupling has been put in place from the very beginning of the Anthropocene— in other words, since the beginning of the Industrial Revolution that the authors of the article qualify as—and this is a crucial point—being European. Instead of simply thinking the Anthropocene from an anthropogenic point of view, we must also consider this new era as a change in governance: the appearance of "global assemblages" transforming state sovereignty merely into one aspect among many others of a "constellation" of various bodies of power (multinational corporations, governmental and nongovernmental organizations). These global assemblages have a tendency to privilege the Global North at the detriment of the Global South, turning the latter into space for the exploitation of raw materials as well as human beings. The article illustrates the variety of ways the "asymmetrical" relations that constitute the global economic system of the Anthropocene—for example, the coffee producers in Papua New Guinea earn an average of fifteen cents an hour for coffee that is sold for twelve dollars per pound at Starbucks; or, another example, the fact that societies living in the Arctic regions have done very little to contribute to the problems of climate change but are at the same time some of the most affected by these climate changes.[10]

In order to break with these asymmetrical relations that lead just as much to the impoverishment of humans as to the deterioration of the ecosystem, the authors of the article put their faith in transnational activist networks fighting to protect local environments from forms of destructive exploitation. As an example, they refer to the ongoing struggles in South America and insist in the way in which certain countries such as Ecuador (but we could also just as easily add Bolivia) have introduced into their constitution rights for nature. Instead of considering nature as a simple exploitable object, nature is considered as a subject that we must respect.[11] A source, and not simply a resource—a "Mother Earth" with rights, one of which would be the power to "regenerate" itself.[12] By starting from a local political point of view, and no longer a viewpoint from the stratosphere, a resilience and a stewardship that leave a place for the alterity of nonhuman worlds can fully be realized. Would this mean that humans would lose a bit of their power? Can the Anthropocene make a place for that which is not simply human or Western?

THE TWO POLITICAL BODIES OF THE ANTHROPOCENE

The only way to create such a place consists in dividing the supposedly unified subject of the Anthropocene in order to shed light on what this false unity strives to conceal. Whereas certain people see nothing more within this concept than Humankind's power over the Earth, the power of humanity as kind practicing a technologically assisted stewardship on a planet that is becoming climatized and integrally anthropoformed, it is essential to provide a place for the existence of a counterpower—the counterpower seen within the variety of activist organizations fighting for their territories. And yet, the entire discourse of the Anthropocene is against this sort of necessary division—the Anthropocene is a political discourse that doesn't say its name, that speaks of Earth stewardship in order to avoid speaking of the various ongoing terrestrial divisions.

Let's return to the way in which the grand narrative of the Anthropocene tells its story, at night around the (nuclear) campfire: "We know, the geo-constructivists say, that we're in midst of the Anthropocene, and we are the first humans to be aware of this." In a certain way, for the geo-constructivists, it's as if the Anthropocene just began, as if *the nominal event* Anthropocene—that is to say, the simple invention of the word—formed in accordance with the *real event*. "Welcome to the Anthropocene!" gives one the impression that we have just entered into this era—it's no longer the era of the great acceleration, but spontaneous generation . . . The mythical grand narrative turns into a complete spectacle, and the Anthropocene turns into a theme park, some kind of Jurassic Park where humans would be the heroes. Particularly intelligent heroes: If we are indeed the first humans to truly understand what is happening to us, and the significant role we are playing in this planetary story, that would mean that the preceding generations understood nothing and had no idea what they were doing to the environment. Hence the idea found in the writings of the geo-constructivists from Carl Sagan to Paul Crutzen: 1) before, humanity changed the environment in an unintentional way; 2) from now on, thanks to the experts of the Anthropocene, humanity can deliberately change the environment. This idea finds one of its major theoretical foundations in the work of sociologist Ulrich Beck. Beck's work postulates a division between a first modernity that is unconscious of its actions, fascinated by the idea of progress, creating the Watt machine out of pure joy and happiness while anchored within a denial of the environment; then there is a second, reflective modernity, capable of taking into consideration the risks and "attachments" between humans and nonhumans—a modernity discovering the fragility of the biosphere. And yet, this temporal division doesn't hold up:

1) On the one hand, the Anthropocene had been actively, deliberately, and consciously installed. As Christophe Bonneuil and Jean-Baptiste Fressoz note, the entrepreneurs of the Industrial Revolution actively shaped the Anthropocene—Saint-Simon was already aware of the fact that the transformation of the globe would ultimately transform it.[13] The development of the coal industry in the United States in the nineteenth century, combined with the more general use of fossil fuels and the rise of the automobile, is not the result of some sort of inexorable progress but of decisions that could have been different than those that were made: The Anthropocene—the two historians say—was a deliberate "thermocene," the fruit of choices concerning energy consumption. The initial choice that was made in regards to energy use was "the choice of fire." And it was this choice that eventually led to the "thermo-industrial civilization."[14] As far as so-called progress is concerned, it would be useful to compare how this choice of thermo-industrialization gave way to what Bonneuil and Fressoz name the "thanatocene": an age of technologically assisted death, reinforced by the massive possibilities for destruction that were created in the twentieth century. War is the pursuit of the Anthropocene by other means. Since the human being is not merely a geological force when creating roads or cities but is a force when destroying them as well, how many deforestations and tactical destructions of territories have been used as motives for war? A myriad of passages abound, leading from thanatological force to urbanological power, and the authors of *The Shock of the Anthropocene* describe the inventions of "brutal technologies" (Paul R. Josephson) that have gone from "military use" to "civilian use"—for example, we can think of the number of

times chemicals used in warfare have been reconverted into pesticides. A destruction with a myriad of faces: the face of war, but also the faces of capitalism and consumerism—to consume materials can also lead to a widespread lethal excess of consumption. In this sense, as Bonneuil and Fressoz explain, the Anthropocene is a "phagocene." This term obviously refers to the idea of a global consumption of planetary resources—an immediate destruction for the satisfaction of needs. But the term also defines a way of producing the nondurable: The capitalist entrepreneurs knew very well what they were doing when they began to spread contempt for any form of recycling; they also know what they are doing when they create the planned obsolescence of objects.

2) On the other hand, there is the "phronocene," an age of prudence (*phronesis*), a sensitivity to the environment, that has accompanied the Watt machine since the beginning of the Industrial Revolution.[15] In fact, since the beginning of the modern period, there has always been "reflexivity": from Charles Fourier proclaiming the material deterioration of the planet, to the twentieth-century scientists who saw the Anthropocene begin to accelerate following the end of World War 2. There has always been a sensitivity to the fragile *circumfusa* (the environmental things) of the eighteenth century. As early as 1770 there was an awareness of the rapport between deforestation and the possibility of climate change, an awareness between the inevitable exhaustion of resources: We knew, and the oppositions made against the "phagocene" were innumerable and unrelenting.

The consequences: Instead of a division *of* modernity, a division between a before and an after, between an initial ignorant modernity that then became knowledgeably informed, what we must consider is a division *within* modernity itself. Instead of a chronological division, we must consider a political division. *Modernity didn't have one head*, a head that was initially naïve and then well-informed, but rather *modernity had two bodies*: one of them deliberately constructed the Anthropocene, starting from precise economic, political, and technological choices whose effects, if not the cause, consisted of completely ignoring the environment and considering it as nonexistent. The other body: a body in fierce opposition to deforestation, a body of petitions and associations formed throughout the nineteenth century denouncing industrial pollution and the maladies resulting from it, a body continually sounding the alarm, having already understood to what extent progress intrinsically generated risks, and to what extent nature and society are inherently interconnected. Such is the double body of the Anthropocene: an insensitive, industrial scorching body and a sensitive, environmental body that is often scorched. If the first body prefers practicing a stewardship of the Earth as viewed from outer space, the second body would like for the Earth stewardship we see emerging to be undertaken within territories inhabited by beings who experience in their flesh and bones what being in the world truly implies.

Anthropocene or Capitalocene?

The great fresco seems to be crumbling before very our eyes: The grand narrative is becoming more and more the figure of a vast lie. Humanity, a unified geological force? It would perhaps be good to remind ourselves that the effects of climate change are the result of specific countries, during specific time periods, and not simply the effects of humanity as such: We shouldn't place the same level of responsibility on countries like Chad or Ghana with countries such as the United States, or those of the European continent and India! In 1900, Great Britain and the United States represented 55 percent of total CO_2 emissions, in 1950, they represented 65 percent of total carbon emissions, and still almost 50 percent in 1980.[16]

In a scathing article, Andreas Malm and Alf Hornborg put forth the critique of the Anthropocene as an ideology:

> Steam engines were not adopted by some natural-born deputies of the human species: by the nature of the social order of things, they could only be installed by the owners of the means of production. A tiny minority even in Britain, this class of people comprised an infinitesimal fraction of the population of *Homo sapiens* in the early 19th century.[17]

For Malm and Hornborg, the Anthropocene is an ideology "by default," not the effect of some sort of political malignancy but the consequence of the fact that the field of studies relative to climate change is dominated by the natural sciences.[18] Humanity can only be defined as a geological force through the effect of a conceptual naturalization or, rather, renaturalization. Studies on climate change prove that social relations determine the natural conditions and denaturalize changes in the climate, thus demonstrating their artificial character; but, Malm and Hornborg argue, within the area of what they call "Anthropocene thinking," "natural scientists extend their world-views to society" and attribute to *Homo sapiens* (*anthropos*, humanity as a geological force, etc.) responsibility for these changes: "Not nature, but human nature—this is the Anthropocene displacement."[19] A displacement the authors also attribute to Chakrabarty, who is accused of succumbing to this naturalism. This naturalism will directly lead not only to an underestimation of class differences but also to the question of race. Not all members of so-called humanity are vulnerable in the same manner, and Malm and Hornborg insist on the way in which this vulnerability directly affects—in vastly different ways—the various white and black neighborhoods in New Orleans after Katrina. Or the way the wealthy or impoverished communities in Haiti or Manhattan are affected after a hurricane, or those in Bangladesh and the Netherlands are affected after a rise in sea waters: "For the foreseeable future—indeed, as long as there are human societies on Earth—there *will* be lifeboats for the rich and privileged."[20]

We claimed that the Anthropocene still has two bodies. But we're taking an additional theoretical step here: In contrast to a *naturalist displacement*, it would be necessary to affirm that humanity is not a global geological agent. Which means two things:

1) Firstly, that humanity is not a *global* geological agent, since there are countries that are different than others (the United States is not Ghana), there are *different class systems* (the rich are not the poor), and there are *racial differences* (African Americans are not placed within the same conditions as white Americans);

2) but this also means that humanity is not a global *geological* agent—in other words, changes in the climate, far from simply expressing an anthropogenic effect, are always *sociogenic*. A conclusion becomes clear: There is no such thing as an age of Humankind, *there is no Anthropocene*. What there is are political, economic, and technological decisions, and social groups, different social bodies that an ideological screen—namely a "naturalizing" screen—has trouble unifying. From this point forward, the mirror of the Anthropocene is cracked—shattered into as many bodies as there are social, economic, and racial divisions. Does this mean that we should get rid of the concept of the Anthropocene? Would shattering this mirror help us to prevent the irresistible installation of the screen of geoengineering?

Before answering this question, we need to recognize a hidden flaw within the theoretical approach that consists of unmasking the effects of "naturalization" specific to the Earth sciences that, according to Malm and Hornborg, orient the dominant discourse of the Anthropocene. In fact, what these two authors don't seem to take into consideration is the fact that the *geo-constructivists don't believe in nature*, whether it be a nature in regard to the Earth or a human nature. What the geo-constructivists believe in is what Jason W. Moore calls "cheap" nature, which is also a kind of degraded or debased nature, as the verb "to cheapen" suggests. And yet, if cheap nature is a kind of debased nature, it's actually *less* than nature: It becomes nature in a diminished form. If it proves necessary to question the term *Anthropocene* and instead focus on the term *Capitalocene*, it's in order to demonstrate that the capitalist economy is first and foremost a "way of organizing nature."[21] The Capitalocene is the mode of the organization of nature that consists in reducing nature to almost nothing at all. Moore claims that this reduction is visible in environmental degradation but first and foremost strives for the *incorporation* of nature within the capitalist economy in the form of a simple factor of production: From now on, nature will work in the service of capitalism. Nature will be considered as nothing more than this service and factor of production.[22]

The quasi-seamless success of this reduction and incorporation of nature into the capitalist economy has been confirmed in the article "Living in the Anthropocene" by the journalist Christian Schwägerl and co-written with Paul Crutzen:

> The long-held barriers between nature and culture are breaking down. It's no longer us against "Nature." Instead, it's we who decide what nature is and what it will be.[23]

In case the message wasn't clear enough, the last sentence of the article rings out as a piercing reminder: "Remember, in this new era, nature is us." But there is nothing at all natural about this "us" that has swallowed the Earth so as to regurgitate it in a new form! It's a colorless "us," genderless and without nationality, that has the capacity for shaping the Earth, as an amorphous planet without a history for *terranthropoforming*. Yes, it's true, on one hand, the geo-constructivist recognizes the relations that unify nature and culture; but on the other hand, he absolutely rejects this space of relations for the benefit of a metaphysical, exceptional, anatural humanity capable of escaping the laws of gravitational force, becoming a cyborg in the heart of space, a creature no longer constrained by breathing or having to concern itself with its material body.[24] Can we really consider such a state of ontological exception as "human nature"? Or rather its opposite? One could argue that the concept of "human nature" was precisely coined so as to conceal the biological and terrestrial dimension of the human condition; but then it's time to get rid of it.

Our goal is not to reject the sociological analysis of the Anthropocene. But this analysis should not lead us, in the name of the necessary critique of "naturalism," to discard Chakrabarty's attempt to rethink humanity as a species— not as a naturalized subject but as a negative one. This is a point we will be returning to in the following chapters: declaring oneself to be postnatural is to completely plunge de facto into the warm embrace of geo-constructivist meaning and forget that the end of the division between nature/culture was effectuated for the benefit of culture, technologies, and human colonization. Thus, we believe it's possible to state that one of the beneficial effects of the concept of the Anthropocene could be, in contrast to an abstract "us" made up of astronaut designers of the spaceship Earth, that it forces us to reconsider the status of the human being: The Anthropocene should be the era where it becomes impossible to not think of ourselves as living beings, the era where "cheap nature" turns out to be an ecological impossibility. However, it has become crystal clear that the astronautical mirror of the Anthropocene and the screen of geoengineering actively prevent this self-awareness, this corporeal commitment to our sociogenic activities, this indisputable recognition of the reality of the Capitalocene.

THE POLITICS OF GEO-CONSTRUCTIVISM AND CLIMATE JUSTICE: THE MINORITARIAN BODIES OF THE ANTHROPOCENE

We have now reached the conclusion of the first part of this book. Geo-constructivist politics is underpinned by what we will call an anaturalist position. The objective of this position is to become capable of getting rid of everything that could induce the least bit of resistance to the reformatting projects of the Earth. However, terrestrial nature is a form of alterity, an outside, a material recalcitrant to the control of engineers: For a geo-constructivist, nature is that which continually risks manifesting itself in an unpredictable nature. To claim that *nature doesn't exist* is, like some sort of magical invocation, to deny the real of this stubborn alterity—the real that risks, through the implementation of climate engineering, making a violent return like the return of the repressed. Furthermore, to claim that the Earth *is not* the Earth but simply a planet without any specific qualities, like some sort of New Continent waiting for the arrival of the geo-constructivist explorers as the new Christopher Columbus, is a way of denying the true alterity, the true singularity of the Earth.[25] As with any geo-constructivist expression, the rejection of terrestrial nature is not a simple description or an ideological veil clothing a discourse; it is performative: The politics of geo-constructivism are founded on the anaturalist axiom that allows geoengineers, as well as any others who share this subjective position—whether they are CEOs investing millions of dollars into research dedicated to a climate shield, journalists, or essayists, such as Mark Lynas who tells us that the human species is "divinely" capable of geoengineering[26]—of *exempting itself from* the possible consequences and futures of its decisions. We have tried to show in considerable detail that this psychological state of exception has been based on a representation of an abstract and *off-planet* humanity.

In a certain way, the whole environmental justice movement is opposed to such a representation. The founding axiom of this movement is nothing more or less than to always grasp the entirety of the ecological and socioeconomic conditions to which each industrial decision (the installation of an incinerator, fracking, the construction of a dam, etc.) may give way.[27] We will therefore follow the thesis put forth by Giovanna Di Chiro: "Environmental struggles are struggles for the reproduction of the social."[28] In other words, and contrary to the "ideological position" of "mainstream environmentalism" founded on the "separation between humans and the natural 'world,'" we must affirm that "people are an integral part of what should be understood as the environment."[29] An attack on the environment is therefore always an attack on the conditions for life, for the survival and reproduction of human life. In this light, it is striking to discern that geo-constructivism seems to reproduce the same division as traditional environmentalism—that is, the division between "humans" and the "natural world"! Nevertheless, *geo-constructivism does not base this division on the necessity for protecting the environment but on the necessity for protecting Humankind*: For the geo-constructivist, it's first and foremost Humankind that must be removed from the recalcitrant materiality of the Earth.

Indeed, a number of characteristics have been removed from this Humankind—that is, erased: "Does the 'anthropos' of the Anthropocene have sexual organs [un sexe], a race, or a gender?," asks Giovanna Di Chiro; and she responds in the negative.[30] It's true, the constructivist politics of the Anthropocene require a humankind capable of confronting anything thanks to its thermo-industrial megaphone, the necessity of geoengineering. In opposition to this politics that only reflects on Earth stewardship from above, environmental justice strives to listen and be attentive to the *minoritarian bodies of the Anthropocene* in order to grant them a voice. In this light, we should recall the violent critique addressed to Chakrabarty by Malm and Hornborg, a critique that we could also extend to all those who consider humanity as a unified, globalized subject: The globalization of the subject *anthropos* is always for the benefit of the dominant body at the detriment of minoritarian bodies. Hence the idea that as long as there will be "human societies," (and by this, we also mean class differences) there will always be "lifeboats" (for the privileged few).

This is geographically undeniable; but it is nevertheless chronologically problematic. Of course, it is the communities of color residing in the Global North and the marginalized people of the Global South who are, for the most part, receiving the lashings of climate change. Sometimes within the silence of the media who undervalue the "environmentalism of the poor" and underrepresent the "slow violence" (Rob Nixon) of the toxic processes that transnational companies export to the Global North.[31] But with the exploitation of extreme forms of energy (fracking, tar sands, strip mining, deep sea mining), including in the Global North, the zones sacrificed on the altar of economic development have a vicious tendency of seeping into the gardens of the privileged. A metonymy of a civilization that made the "choice of fire," fracking carbonizes everything it touches: Rocks, mountain summits, gardens, stubborn indigenous peoples, *everything* that hinders access to fossil fuels must be destroyed.

"We're all in the same sinking boat," proclaims Deeohn Ferris (a public figure of the environmental justice movement), who also adds, "only people of color are closest to the hole."[32]

One of the most crucial stakes of contemporary political ecology will be to solder together questions regarding the environment with those regarding questions of race—as the members of the Midwest Compass Group have recently written, "Political ecology begins when we say: Black Lives Matter."[33] This welding together will allow for different types of minoritarian bodies of the Anthropocene to coalesce, these bodies that are the most exposed to the harsh realities of the Anthropocene and the other minoritarian bodies to come who must understand that they will end up in the same position once human societies no longer persist except under conditions that are becoming more and more unlivable—more and more deprived of what we can really call "a world."

PART II

THE FUTURE OF ECO-CONSTRUCTIVISM

From Resilience to Accelerationism

Thank God, Nature is going to die.
—Bruno Latour, *Politics of Nature: How to Bring Sciences into Democracy*

Never mind terraforming Mars; we already live on a terraformed Earth.
—Stewart Brand, *Whole Earth Discipline: An Ecopragmatist Manifesto*

TURBULENCE, RESILIENCE, DISTANCE

On August 15, 1971, the gold standard, the conversion of the dollar into gold, was suspended. Two years later, the currency exchange became "floating," which meant that from now on, the rate of exchange would be determined by the state of market fluctuations. And it was in this manner that the Bretton Woods Agreement, which had regulated the international financial system since 1944, came to an end. What happened at that time with regard to important economic data is something that obviously has an important bearing on how we largely structure our present—such a fluctuation of the exchange rate has largely favored speculation on currency and a disconnection of the speculative sphere, its autonomization from the so-called "real" economy. However, we shouldn't convert this economic data too quickly into a hastily formed explanation for what should be more properly described as a major epistemic change, even a change in civilization, a major upheaval in the way we think about science, politics, the economy, as well as ecology and the environment.

To put it in synthetic terms, the paradigm change that took place throughout the 1970s consisted of:

1) Ontologically establishing our civilization and all its aspects around the concept of turbulence (chaos, disorder, and disturbances);

2) Installing, on top of this ontological collapse, a political economy of turbulence capable of managing the "risks" that are necessarily induced by this ontological situation: Such a political economy requires of its subjects to *adapt themselves* to this ontological chaos and its *programmed uncertainty*.

The eco-constructivism that this present chapter will attempt to describe was born out of this context. It took shape by way of forging its own concept of resilience; it adorned itself with modernist, pragmatist, post-preservationist, post-environmentalist attire, always ready to change costumes and customs since the world itself is always endlessly changing; it built its reputation on the "political ecology" of Bruno Latour: his theory of attachments and of the actor as "what is *made to* act";[1] eco-constructivism strives to find some sort of future through accelerationist approaches, or even transhumanism, that seem to only speak of ecology in order to better dissolve it. Is eco-constructivism digging the grave of ecology?

In contrast to those who would like to somewhat hastily bury the environmental movement, who only swear by virtues of a technologically unbridled capitalism, who believe that the Anthropocene is an opportunity for humanity, who sing the praises of our indivisible attachments, we must propose some sort of *detachment*. Not in the form of a withdrawal into some kind of imaginary fortress, not some sort of technological decoupling that would place us into an orbit around a reformatted Earth, but a *distance* within the world without which no politics whatsoever is possible. Except, of course, the automatic politics of limitless development and the delights of a *democracy of the economy*—if this is how we want to name the naive subjection of the political sphere to the economic sphere.

CHAPTER 5

AN ECOLOGY OF RESILIENCE

The Political Economy of Turbulence

The ontological axiom of the paradigm of turbulence can be described in the following way: Nothing is ontologically anchored; nothing is definitively stable; nothing is assured a priori. Such an ontological thesis bears the resemblance to a certain déjà vu: Is this not what Nietzsche named the "death of God," a theme we evoked in the beginning of this present work by citing *The Gay Science*, published in 1882? Is this not also the declaration made by Marx and Engels sketched out some ten years prior to Nietzsche, in the *Manifesto of the Communist Party* (1847), where they identify "this perpetual uprooting of production, this constant unrest of every social system, this perpetual agitation and insecurity" as the characteristics of an era where "everything which had solidity and permanence goes up in smoke"?[1] What distinguishes these defining characteristics of the nineteenth century from those of our current era? The fact that, today, destabilization is no longer simply an effect of a social class (the bourgeoisie), it has become ontological: *Destabilization has become the established order*. Destabilization is not simply the effect of actions taken by the bourgeoisie but has become an active cause. It's as if the world itself has become a zone of various kinds of turbulence: from the turbulence of the financial sphere to the "great aerial ocean" (to use Alfred Russel Wallace's metaphor to describe the atmosphere).

From this point forward, the *chaos sciences* that model this paradigm will bear the responsibility of explaining the relationship between order and disorder, the genesis that explains the passage from the nonorganized to the organized, from ontological instability to the relative stability that ordinary experience continues to verify—of course everything changes, but the Earth keeps on spinning within a solar system that, up until now, hasn't burst into a thousand pieces. Published in 1979 with the title *Nouvelle Alliance. Metamorphose de la science*, the book by the physicist Ilya Prigogine and philosopher

Isabelle Stengers will quickly be published in English as *Order out of Chaos*. Counter to determinist sciences, Prigogine and Stengers describe our world as a universe that is "irreducibly aleatory" in which reversibility and determinism "become figures of particular cases."[2] Instead of starting off by way of thinking equilibrium and then disturbances, their physics of nonequilibrium posits 1) first there are fluctuations, improbable bifurcations, which then lead to search for, 2) in a second phase, how "dissipative structures," systems "far from the equilibrium," become organized and, more specifically self-organized.[3] We should be clear here that, for these thinkers, it's not a question of fetishizing chaos or promoting it in some kind of romantic way, wherein one would strive to maintain one's footing within the disordered chaos as if in a kind of ecstatic rapture; rather, it would be a way of understanding the stability or metastability of these systems that are far from equilibrium. Prigogine and Stengers's approach is one of the modalities of "chaos theory," which studies the behavior of systems that are extremely sensitive to their initial conditions. To the extent that the slightest disturbance on the ontological plane can give way to phenomena—by changing scale, by passing from one plane to another, from the microscopic level to the macroscopic level—to which no determinist straight line would have led. These systems that are difficult to predict—such as the atmosphere—are called "nonlinear" (hence, the well-known "butterfly effect" explaining how a local disturbance can effect a global modification).[4]

The idea of a turbulent world that we cannot predict became widespread within the field of economics beginning in the 1970s.[5] According to this approach, economic systems must be considered as "nonlinear," sensitive to local variations—variations of affects, beliefs, and ideas. But we should be careful not to commit the error that would consist of putting on the same plane the various ways in which, on one hand, various sciences and theories in physics and, on the other hand, the techniques of government and economical-political management relate to the uncertainty of chaotic systems. It's certainly true that these sciences and political-economical techniques regularly exchange their signifiers (such as the term *resilience*, which we will examine later). But we should highlight the difference between them: The problem for physicists is to determine the laws that regulate the *immanent* passage from disorder to order as well as maintain a structure that is "far from equilibrium" (Prigogine); but the problem of nation-states, of transnational firms, of the *think tanks* who supply ideas to them, is that of proposing the *transcendent* ways for preventing pure chaos—that is, to speak plainly, to prevent political dissidence and insurrectional "risks." By "transcendent ways," we mean the act of *deciding*, of choosing this or that politics. To put it another way, they must choose between attempting (or not attempting) to stabilize a political system or attempting to turn the instability of a financial market into a profit for its own benefit: This is a mode of government or "governance." It is a result of a kind of "management" that is itself *contingent*.

In order to elucidate this point and its consequences, we must understand that the political economy of turbulence is founded upon the following two ideas:

1. *Turbulence is inevitable*—this is the ontological axiom of the world as turbulence, or the *axiom of turbulence* we have been examining since the beginning of this section.

2. As a consequence, we must learn to *manage this risk*—that is, we must learn how to adapt. As Melinda Cooper writes, within such a paradigm, "turbulence cannot be avoided; it can merely be managed."[6]

If "risk management" has its origins in the 1950s, it's the collapse of the Bretton Woods system resulting in the convertibility of monetary currency in 1973 that led to our current understanding of the meaning of "risk management." It will no longer be a question of simply assuring ourselves against catastrophes, as if these catastrophes were abnormal events breaking with daily life, as if they were coming from the outside. From now on, risks will be integrated, normalized, treated as ordinary and recurrent facts. Welcome to what the sociologist Ulrich Beck named, at the end of the 1980s, the "risk society." Within the economic domain, from then on we speak of "financial risk management" and "adaptive management."[7] To put it another way, the political economy of turbulence is defined as a synthesis between chaos and adaptation. Yet between the axiom of turbulence and the adaptive consequence, there is no logical passage whatsoever, there is no obligatory passage. The political economy of turbulence was not a potentiality preinscribed in Prigogine's nonlinear physics, and our present goal is not to propose an "ideological" critique of the book, *Order out of Chaos*, which he coauthored with Isabelle Stengers. If the famous ultraliberal economist, Friedrich Hayek, known for his defense of a self-organizing market that would be so efficient that it would surpass any state control, cites Prigogine as one of his main sources of thinkers who allowed him to contest the models based on equilibrium that were favored by the neoclassical economists, this support for his theory is not enough for explaining why Hayek pleads for us to denounce "the hubris of predictive modeling in the face of unknowable complexity" of the "social and ecological systems [that] will evolve most productively once liberated from the counter-evolutionary control of the interventionist state."[8] Here we uncover the paradoxical point: The political economy of turbulence denies the contingency of its ideological-political option, even though it posits this structure of contingency as the ontological structure of the world! The political economy of turbulence compels everything and everyone to adapt to a chaos that is essentialized, turned into a fatality in front of which humans and nonhumans, "natural" or anthropized environments, must become resilient.[9]

INSTABILITY AND RESILIENCE

Resilience is one of the favorite words used in the environmental sciences. It's also a term used frequently in the fields of economics, political science, and psychology, each time one attempts to understand why and how the turbulence of the world—environmental changes, hurricanes, terrorist attacks, economic crises—doesn't lead to the pure and simple destruction of the individuals, social systems, or ecosystems in question but rather to their persistence—even if this simply means their minimal survival, of the transformation of the modes of life or the biotic environment.[10] In the area of ecological sciences, this notion has been shaped and sculpted by Crawford S. Holling, whose work has had and continues to have an enormous influence on the field of environmental studies and the institutions that have brought the term *resilience* into power, such as the Resilience Alliance (founded in 1999 by Holling) or the Stockholm Resilience Center (founded in 2007), the latter defining the term *resilience* on its website as "the capacity to deal with complexity and change" and promoting a science "for the sustainable co-evolution of human civilizations with the biosphere."[11] *Turbulence, change,* and *development*: How are all these terms linked within and by an ecology of resilience, which is one of the primitive components of eco-constructivism?

In a somewhat troubling coincidence, Holling's founding text, "Resilience and Stability of Ecological Systems," was published in the same year as the end of the Bretton Woods Agreement (1973). Holling targets a specific kind of ecology that up until then had a reserved position of dominance: an ecology of order, of harmony, of a nature in an "equilibrium state,"[12] an ecology submitted to what Donald Worster calls "Clement's Paradigm." Indeed, the ecologist Frederic Clements (1874–1945) saw the succession of plants colonizing a territory as the slow and progressive formation of an order leading to what he called a "climax"—that is to say, the state in which the vegetation is the best suited and adapted to its environmental conditions. This state of climax was supposed to give birth to a "superorganism" wherein the various parts of the vegetation form a solidly constituted whole from then on capable of self-organizing and maintaining itself at a high level of stability.[13] It's this paradigm that Holling destroys or, rather, more precisely, *reverses*: Instability is there first, and the true question becomes knowing what ecosystems do in order to "persist." The answer to this question is given thanks to the concept of resilience, which Holling defines as the "high capacity" that "natural systems" have for "absorbing change without dramatically altering."[14] Actually, Holling does nothing more than simply reverse the paradigm of classical ecology. He changes the function of the terms constituting it: Instability is not in opposition to the possibility of persistence but rather constitutes its condition of possibility. In fact, the promoters of the concept of resilience maintain that systems have reserves that enable them to adapt and not disappear in the event of important changes within the environmental conditions *because* they are unstable. A contrario, the more a system is stable the fewer chances it has for being able to generate an adapted response in the event of change—in other words, the system is less resilient. Instability is therefore the source itself of resilience, understood as the capacity for changing without disappearing.

However, such a definition for the term *resilience* might be surprising. Since the word was first used by engineers for defining the capacity of a material to return, after some kind of deformation (for example, due to heat), to its initial state. And yet, the entire theory of resilience is opposed to such an eventuality. In a world where everything is turbulent, there is no possible return whatsoever. It is, literally, move or perish, change or die. But to change, to move when everything around you is changing, signifies evolving—that is to say, learning the new way in which the dominant political economy will attempt to impose itself as the norm. Consequently, we must distinguish between two types of resilience: on the one hand, engineering resilience, which considers it possible, thanks to technological mastery, to control the environment and predict the consequences of our actions; and on the other hand, the resilience induced by the ecology of chaos, this "new ecology" that is tasked with understanding "the dynamics of non-equilibrium and the adaptability of ecological systems to vast and uncertain changes."[15] Ecological resilience turns ecological systems into mega-chameleons capable of changing forms in a formless world . . .

In no way is this description of resilience an exaggeration. Here is how Holling describes the world of resilience in an article he cowrote in 2002: a world of "complex non-linear relations between entities under continuous change" confronting "discontinuities" and "uncertainty," these latter two manifesting themselves by way of "synergetic stresses" and "shocks."[16] In such a world, resilience is "the capacity to buffer change, learn and develop."[17] We then can understand that the simple act of passively withstanding shocks is not enough to explain adaptation. Whether it is natural, social, or socio-natural, the system must be understood as being capable of learning and developing in order to respond to new situations. But to what degree can a system learn? To what degree can it adapt? Holling had this problem in mind already in 1973:

"This resilient character has its limits," and when these limits are exceeded, "the system rapidly changes to another condition."[18] Very well then. But what does change mean when everything changes? What does learning to change mean if changing means changing into something completely different—different in relation to what sort of referent, if the initial referent is chaos, if the idea itself of an original equilibrium no longer has any meaning whatsoever? The ecology of resilience has so completely accepted the axiom of turbulence that it finds itself in the situation of being ontologically incapable of giving an account of the turbulence that nourishes it.

FROM RESILIENCE TO PANARCHY

Far from being purely speculative, these problems will end up being examined by the popes of the ecology of resilience, Lance H. Gunderson and Crawford S. Holling, in a book published in 2002 called *Panarchy: Understanding Transformations in Human and Natural Systems*. The preface and the introduction immediately focus on the problem of problems: change and, more specifically, "the interplay between change and persistence, between the predictable and the unpredictable."[19] In fact, the book does not contest the concept of resilience: Yes, nature is resilient. It is capable of self-organization and of inducing biotic variation adequately responding to new situations. Nature is therefore neither static nor in equilibrium. All this is very certain. But this resilient nature is only a partial representation—what the authors even call a "myth"—of what nature is, since nature is also "nature evolving." But, couldn't we say that this idea of evolving is already contained within the concept of resilience? No, to the extent that nature evolving is "a view of abrupt and transforming change."[20] And this is what allows us to identify two problems:

1. Such a change can lead not only to a transformation but to a destruction.

2. Resilience is not necessarily a quality. In order to explain this point, Gunderson and Holling provide an example of a sinister type of resilience . . . of the bureaucratic dictatorship of the ex-Soviet Union, in specifying: "Resilience is not an ideal in itself."[21] In a certain way, the concept of resilience is too static! In and of itself, it doesn't allow for the integration of other powers of destructive creation that sometimes require the death of the system itself.

Hence the necessity to make the schema of analysis more complex, to demonstrate, in a certain way, that complex systems are even more complex than they appear . . . What must be taken into consideration is *the "cycle of change"* of which resilience is only one aspect. This cycle experiences a phase of exploitation (labeled "r") where the ecosystem establishes itself and develops; then there is a phase of conservation ("K"), in other words, the capitalization

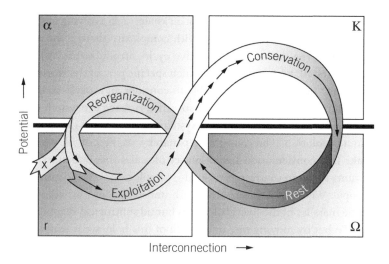

FIGURE 2. Adaptive Cycle. Reproduced by permission from Lance H. Gunderson and Crawford S. Holling, eds., *Panarchy: Understanding Transformations in Human and Natural Systems* (Washington: Island Press, 2002).

of resources; a phase where the system slackens or becomes less strict ("Ω"), disintegrates, a phase of "creative destruction" (an equation borrowed from the economist Schumpeter, who used the term to speak about capitalism . . .); then there is a phase of reorganization, of renewal ("α"). The becoming of these four phases is conditioned by the "potential" of the system—that is to say, the resources thanks to which a creative adaptation will be possible (the biomass of an ecosystem and all of its nutritive substances)—and its degree of internal "connectivity": The more a system is connected, the more it is capable of self-conservation. But this quality also contains its opposite effect: Too much connectivity inevitably implies a loss of plasticity. A hyper-consistent system, closed in on itself, solidified by a network of internal relations, will not be able to adapt to a change in environment and will collapse. The change, specific to the phase of reorganization, will only be possible if the chaos that is unleashed at the moment of self-destruction can be used as a source for a new organization—here again we find this idea of *an order out of chaos* specific to the axiom of turbulence. The adaptive cycle will then be represented as a Moebius strip, a figure eight that is actually a unilateral surface (a topological surface with only one side endlessly becoming the other), in order to show that all of the phases are not added onto each other but constitute the same unity without recto and verso (i.e., *without any absolute rupture*: Everything happens on the same continuous ecosystemic surface).

All right then, resilience is nothing but an aspect of an evolving global cycle. But, and this is one of the problems with complexity theory, the model of integration that the notion of an adaptive cycle introduces proves to be insufficient: After all, which cycle, and which specific part of the world, are we talking about? In truth, it is impossible to consider an ecosystem in an isolated manner, as the evolution of this so-called isolated ecosystem can only be falsely separated from the other ecosystems and the social groups that play a role in its evolution. The concept of "panarchy" is, first of all, what allows us to take into consideration a multiplicity of scales that one must consider if one wants to begin to grasp the—real—complexity of the adaptive cycles. An example: We must task ourselves with understanding this alligator hole within the marshes of the Everglades, this humid subtropical region of southern Florida; but by moving one mile north of this hole, we uncover a vastly different landscape, one made up not only of marshes but of large masses of algae, and humid prairies welcoming a number of invertebrates; if we move up a bit further from there, we will discover and have to take into account the entire socio-ecosystem of southern Florida, with its concentrated zones of populated areas surrounding Miami; and then, if we go a bit higher up, the entire Florida peninsula; then the bioregion of the Gulf of Mexico, and finally, the Earth.[22] So let's first provide a definition for *panarchy*: *pan* in Greek means totality, and panarchy is the desire to take into account all the different scales of the Earth's ecosystems but without any sort of hierarchy (i.e., without any kind of predominance of one scale over another). What the concept of panarchy invites us to consider is that there are different scales of the Earth's ecosystems. So we can understand them, measure them, and decide on the desired or undesired changes to them—to then ask ourselves: Should we destroy them or conserve them?

Everything changes, this is the message presented by the Greek god Pan, who shares his name with the concept we're studying; but change is a flux that integrates conservation as a moment of change as well, a slowing-down, and that sometimes becomes a lethal immobilization [fixation]. However, we should never forget that there are *several* changes, *several* scales, *several* temporalities: The adept of panarchy must be capable of modulating her gaze in terms of the pursued objectives. She must take into consideration the system constituted by all these various figure eights, all of the various socio-ecosystems that, however interrelated they may be, must be read and interpreted through very distinct scales of time and space. Only this multiscalar reading can allow one to know whether or not the "Ω" phase—the release phase—must be avoided. For in the end, only this phase, that is sometimes rather painful, can allow for a change or an improvement of the system. We should finally add that the "α" phase—the renewal phase—is not in the least bit guaranteed: The system can, all by itself, collapse without changing—that is, simply reach its final degree of change, which is pure and simple destruction. Pan is also familiar with death.

FOR A REDEMPTIVE WITHDRAWAL OF PAN

A priori, is there anything more beautiful than such a vision of ecology? Having become panarchic, an ecology of resilience is an ecology open to surprises to the second power: Not only do unpredictable things happen, but the consequences of these events are themselves unpredictable.[23] Everything changes and also modifies our way of reacting to these changes. As we have already mentioned, this situation leads to some difficulties in regard to questions of analysis and forecasting. The postmodern Pan that presides over the fate of the political economy of turbulence is a god that has become rather cryptic, a god so complex that it has become perplexing, troubling, and incapable of clearly following a straight path—unless it's the one imposed upon it by the democracy of the economy and its geo-capitalism.

What, in fact, is the strategy of the ecologists of resilience? Confronted with the conceptual difficulty posited by the axiom of turbulence, their strategy is one of integration. The integration of the difference of scales, of time and space, the integration of natural and social systems, the integration of diverse representations of nature, etc. The integration of the integration of the *whole*— as we know, this is also a definition of the word *pan*. But what does the integration of everything actually mean, if not for it to become incomprehensible? Or perhaps just as ridiculous as the person who would like to construct a map that would be as big and as realistically detailed as the territory it would attempt to reflect? And here we can identify the differences and commonalities between geo-constructivism and the economy of resilience:

1) A priori, as we have seen, the ecologists of resilience are against the vision put forth by the geoengineers: The geoengineers believe that resilience is the return to a state of normality after a shock. A confounding naivety if there is no such thing as a state of normality! Even more serious, to consider humanity as "detached" from nature and aspiring to "control" it can only lead to an increase in "vulnerability" and an "erosion of resilience."[24] A perfect illustration of this analysis: A chemical shield would render the atmosphere entirely dependent on some sort of technology—in other words, to take back up schema of the "evolving cycle," it would stabilize and place the atmosphere into a static holding pattern within a phase of conservation ("K"), a phase that, from the very fact of the atmosphere's dependence on a technology, would make it absolutely impossible to adapt to an unpredictable situation. And in this case,

technology would lead to the drying up of the creative potential of turbulence within the atmosphere.

2) However, in reading Gunderson and Holling's book, *Panarchy*, we can find the following statement: "Humanity has yet to become the terraformers at a planetary scale suggested in science fiction."[25] How is it that, in spite of their suspicion regarding the geo-constructivist discourse of engineers, the ecologists of resilience share with them one of the same fantasies? Perhaps it comes from the desire for wanting to integrate *everything*. When all is said and done, the figure-eight diagram of an "adaptive cycle" resembles a topological trap where we risk losing any sort of bearing endlessly moving, without being capable of perceiving it, from one position to its opposite. In a certain way, the ecologists of resilience suffer from an ailment that all eco-constructivists suffer from: *the illness of a lack of separation*. Incapable of entertaining the slightest withdrawal from a world of complexity, within a position of permanent change, endlessly being tossed about by the various kinds of turbulence affecting each aspect of the world, the ecologists of resilience have turned Pan into a stifling totality, into a kind of suffocating box; not an empty box, like that of the geo-constructivists, but one bubbling over with convulsions.

Perhaps they have forgotten another meaning of the word *resilience*: Etymologically, the term is derived from the Latin *resilire*, which means to "bounce back, to rebound, to withdraw," and within juridical language, "to renounce, to retract."[26] As the founding editors of the journal *Resilience*, Stephanie LeMenager and Stephanie Foote, wrote back in 2014, the term can be understood as the act of "drawing back, to distance oneself from an undertaking"—for example, to distance ourselves from something we find repugnant:

> We appreciate that the populist notion of resilience as "bounce-back" includes disgust at the way things are, a necessary self-distancing from the normative.[27]

Without such a distance, it is simply impossible to be resilient in the sense proposed by Holling and his colleagues. Without distance, how can we know how to adjust ourselves to unpredictable shocks? Abolishing one's capacity to separate oneself from a situation, abolishing one's ability to break away from the links of a network, is to condemn the candidates of resilience to adapt themselves for survival without ever being able to question the fact that the turbulence of which they are victims was perhaps knowingly constructed— such was the case with the "turbulences" from Fukushima or the various "turbulences" that bore the name [hurricane] Katrina. When the capacity for self-distancing is denied, the call for adapting is indiscernible from a sacrifice at the altar of the political economy of turbulence.

CHAPTER 6

THE EXTRAPLANETARY ENVIRONMENT OF THE ECOMODERNISTS

ECOMODERNISM VS. OLD ENVIRONMENTALISM

The ecology of resilience has greatly affected those who, since the dawning of the third millennium, have presented themselves as ecomodernists, green modernists, eco-pragmatists, or even post-environmentalists. All of these promoters of a so-called new ecological thought inscribe themselves within the wake of a paradigm shift that we have identified as starting with the axiom of turbulence (the idea according to which the world is ontologically and irremediably unstable) and of its practical consequences (the necessity of adapting to change, of managing the consequences of our actions as well as those changes arising from the ecosphere as "surprises," and the flexibility that such a resilience imposes). But if this ecology in search of its legitimacy defines itself as modern, pragmatic, or post-environmentalist, this is not simply in order to recognize this paradigm shift and to be able to define its *identity*, it's in order to delineate its ideological *opposition* with what it calls old environmentalism.[1] To understand this merciless war is to take into account what could become, in the years to come, the new face of ecology or environmentalism—namely, as we will soon understand, its burial.

Ecomodernism finds its origin within the project of "ecological modernization" that was born in Germany in the 1980s. At the heart of this project is the idea that it is possible to integrate all the environmental concerns found within capitalism without modifying its structure—that is, without renouncing development and progress.[2] This project of modernization that sometimes bears the name of the "green industrial revolution" claims to be profoundly optimistic, in contrast to any sort of catastrophist type of discourse: The ecological question is seen as an opportunity for the capitalist economy[3] (i.e., as an opportunity for a new moment of primitive accumulation, to borrow from Marx) that an antiestablishment political movement could do nothing more but hinder. It is then possible to consider ecomodernism as the discourse whose major function will be liquidating any possible hindrance to industrial development established on taking into consideration environmental concerns—it's in part thanks to ecomodernism that geo-constructivism was able to become hypermodern. In order to corroborate this approach, all one has to do is read the claims made by the most recent representatives of ecomodernism:

> Ecomodernism . . . offers a positive vision of our environmental future, rejects Romantic ideas about nature as unscientific and reactionary, and embraces advanced technologies, including taboo ones, like nuclear power and genetically modified organisms, as necessary to reducing humankind's ecological footprint.[4]

So there will be a profoundly reactionary, technophobic, catastrophist environmentalist movement believing in the virtues of a "nature" that knows what to do—in other words, dividing the question into a nature that we must protect and a humanity guilty of its diabolical destruction: This movement would be antimodern as the Romantics were assumed to have been, rejecting progress and hampering the future. Counter to this environmentalism, a new movement would form, devoid of ideology and in this sense "pragmatic," tasking itself with creating technologies that are environmentally friendly (sic)— such as nuclear power,[5] GMOs, and fracking[6]—and rejecting the division between nature and technology, nonhumans and humans: an ecomodernist movement—a movement that maintains ecology and environmentalism will only have a future once they have eschewed any sort of catastrophism and joyously embraced specific technologies and the economies that underpin them.

Among the figureheads of this new division between the "ancients" and the "modernists," we must mention Ted Nordhaus and Michael Shellenberger, the founders of the Breakthrough Institute who—and this is a point we will come back to later—were able to welcome Bruno Latour as well as a Pascal Bruckner as senior fellows, the latter being well known for his hatred of ecology and environmentalism.[7] In 2004, Nordhaus and Shellenberger published an essay titled "The Death of Environmentalism," in which they criticize the flimsy vision of environmentalists obsessed with this "thing" called the "environment." However, the two authors claim that the environment as some natural thing cut off from humans doesn't exist. In only dedicating themselves to their own specific "interests," environmentalists condemn themselves to not connecting their causes with those of industry and trade unions. Refusing to consider the environment as something separate should lead to connecting interests and creating new alliances.[8] Several years later, Nordhaus and Shellenberger edited a collection of essays titled *Love Your Monsters: Postenvironmentalism and the Anthropocene*. Because of their nonmodern approach to nature, Nordhaus and Shellenberger argue, environmentalists are unable to reduce the carbon footprint of human beings in a world with "seven to ten billion souls seeking to live energy-rich modern lives."[9] The only solution they posit is that "we must once and for all embrace human power, technology, and the larger process of modernization."[10] Fundamentally, Nordhaus and Shellenberger maintain that there are no natural limits for either human beings or the rest of so-called nature. For human beings have always lived in relation with a technological universe that has also in turn made them. Instead of stupidly refusing technologies, we must use them as the means for "saving" ourselves.

Saving the planet does not mean preserving it from all contact with human beings. In fact, such a preservation is impossible: How does one determine the natural state of nature? A natural state can only be determined within a determined state in the past, since ecosystems are constantly changing. The connections here between post-environmentalism and the ecology of resilience are obvious: The world is endlessly changing, turbulent, so environmentalists are incorrect, they are reactionary and dangerously romantic when they seek to protect ecosystems under the pretext of an equilibrium that has never existed and never will. This will be all the more true during the era of the Anthropocene: How does one even begin to determine a natural state that is preserved from human actions when human beings have so thoroughly modified the ecosphere that they themselves have become a major geological force? One must be, resolutely, post-preservationist.[11] Therefore, saving the planet can only mean one thing, and this is one of the leitmotifs of post-environmentalism: *Intervene even more*—in other words, "creating and re-creating [the Earth] again and again for as long as humans inhabit it."[12] This increase in intervention should in no way arouse fear. Quite the opposite: It's fear that we must do away with, along with the other "apocalyptic fears of ecological collapse" that turn environmentalism into an "ecotheology."[13] In contrast to this archaic religion, we must posit a "theology of modernization" that envisions technology as something "morally humane and sacred."

A PLANET OF NO RETURN

Within the post-environmentalist and ecomodernist discourse, the Anthropocene is the confirmation of the necessity of getting rid of the old environmentalists who don't understand that the world is hyperconnected, that humans are everywhere a part of the landscape, and that anthropization materially renders the idea itself of preservation impossible. This thesis is deployed perfectly by Erle Ellis, a geography professor studying environmental systems. For Ellis, environmentalists are wrong in continuing to support the idea of overpopulation—and he claims without hesitation that there really is no such thing as a human carrying capacity: There are no natural limits, Ellis tells us in an article published in the *New York Times* from 2013, "the environment will be what we make it."[14] Let's cite a long passage from another article by Ellis, titled "The Planet of No Return: Human Resilience on an Artificial Earth," which is found in the journal of the Breakthrough Institute:

> The Earth we have inherited from our ancestors is now our responsibility. It is not natural limits that will determine whether this planet will sustain a robust measure of its evolutionary inheritance into the future. Our powers may yet exceed our ability to manage them, but there is no alternative except to shoulder the mantle of planetary stewardship. A good, or at least a better, Anthropocene is within our grasp. Creating that future will mean going beyond fears of transgressing natural limits and nostalgic hopes of returning to some pastoral or pristine era. Most of all, we must not see the Anthropocene as a crisis, but as the beginning of a new geological epoch ripe with human-directed opportunity.[15]

This passage and, in a more general way, Ellis's entire article are interesting for several reasons: 1) For one thing, the article confirms its relation with the ecology of resilience through its subtitle ("Human Resilience on an Artificial Earth") and by the emphasis it places on the question of change; 2) the article then describes its position as modern and, more specifically, ecomodern, by way of refusing any sort of position that could be judged as backward-looking: the fear of transgression, the false hope in any possible return to some sort of past, etc. Being modern means looking straight ahead, and in this sense, continuing to believe in progress, on the condition of being saved; 3) if the planet is at a point of no return, it's not simply because we cannot go back in time but because we also already live on an "artificial Earth"—namely, an Earth that has already been definitely anthropized. And here we will demonstrate to what extent Ellis's ecomodernism, just like Nordhaus and Shellenberger's post-environmentalism, is at the very least compatible with geo-constructivism:

—Of course, as with all environmental or ecological discourse, ecomodernism and post-environmentalism insist on the fact that *the environment is not separate from human beings*: One of the flaws of the old version of environmentalism would precisely consist of maintaining this divide.

—But, according to what we call the paradox of the Anthropocene, these same discourses need to be able to position humanity as being *off-planet*, cut off from the Earth, in order to then "create" the future that we want, as this new geological era offers up new "possibilities." We, humans, we will be capable of choosing a "good" Anthropocene, of granting to the environment the form that we would like for it to have, so as to be capable of "re-stitching and refashioning the coat of Earth stewardship." Yes, this latter expression cannot be mentioned without giving off a feeling of déjà vu! It's true—ecomodernism can find one of its greatest spokespersons within geo-constructivism: Ellis considers the Earth as an empty box, an artificial thing subjected to our technological operations.

But suddenly, the title of the article—"A Planet of No Return"—takes on an entirely different meaning: the enthusiasm of the ecomodernists, their refusal to admit or consider, at all costs, the natural limits of existence, their complete denial of the idea of any kind of planetary "crisis." Does this position not prevent them from coming back down to Earth? Does their position not, in fact, condemn them to remain at the point of "no return," hovering within an ideal stratosphere (i.e., within the environment of their extraplanetary ideas)? Do they not, in fact, project onto the old version of environmentalism their own abusive separation between, on the one hand, an environment of euphoric ideas of the benevolent actions of the Anthropocene and, on the other hand, the less enthusiastic reality of a world that is completely and utterly wrecked?

IS POST-ENVIRONMENTALISM PRE-ANTHROPOCENIC?

Let's try to avoid several misinterpretations. First of all, it's not a question of denying that nature is in a state of perpetual change; nor that the act of preserving an environment can only happen by way of some sort of human action or intervention that we must nevertheless temper. In this light, it is perfectly true to affirm that the situation of the ecosphere is conditioned by way of the decisions that human societies will end up making. But certain decisions can also consist of *not* entailing or unleashing a given experience, such as not drilling within a protected zone, or some other geoclimatic operation. Within the ecomodernist optic, every decision must inevitably lead to achieving something, to a production, to producing something at all costs, at all times; but here what is as stake, as we will develop further along in our present discussion, is not a neutral option but an ideological one: the stakes of the ideology of modernity, this imperative that makes the decision to abstain from entertaining all possibilities of achievement or production impossible.

Furthermore, it's not about being against all forms of technology. It is a certitude that humanity is unthinkable without technics [technique], which is precisely a part of what we call humanity—if we call technics, in a more broad sense, the use of tools, the use of machines, as well as all the physical and mental qualities required for the use and conception of these tools and machines.[16] But, on one hand, *the ecomodernists reduce technics to technologies* by insisting on the quasi-magical qualities of the latter for *automatically* saving us. On the other hand, certain ecomodernists are curiously ignorant of technological evolution as if nothing has changed since the Paleolithic era. Ellis writes, without the least bit of hesitation:

> Since prehistory, human populations have used technologies and engineered ecosystems to sustain populations well beyond the capabilities of unaltered "natural" ecosystems.[17]

This belief in the continuity of history is in perfect contradiction with the definition of the Anthropocene: Whereas Ellis sees nothing but some form of historical continuity, the Anthropocene means the arrival into a new era. As for Nordhaus and Shellenberger, they argue that, from the Neolithic period to our present day, the way that human beings have shaped "nonhuman nature" has not changed in "kind" but in "scale."[18] Here again we encounter the same problem as with Ellis: If a change in scale in no way provokes a qualitative change whatsoever, no change at all in "kind," then why are they speaking about the Anthropocene? Such is the trap from which the post-environmentalists are unable to extract themselves: On one hand, they want to profit from the Anthropocene-event so as to launch their grand program of modernization, so they have to insist on the *rupture* created by the Anthropocene in order to then get rid of environmentalism; and on the other hand, they have to deny or minimize the extent of this rupture by insisting on the *continuity* of eras, under penalty of having to recognize the necessity for profoundly changing our mode of civilization and its values that have led it to where we find ourselves today. Subsequently, it becomes entirely legitimate to be fearful in regard to a future that would be shaped by post-environmentalists. Yes, it's perhaps a bit too soon to become a *post*-environmentalist and to consider the classical environmental movement as "outmoded."[19] On the contrary, it may very well be that the tacit or fervent geo-constructivism of the post-environmentalist or ecomodernist "ecologists" is merely the sign of an inability to comprehend the fundamental philosophical rupture to which the Anthropocene must expose us.

This becomes patently obvious in reading *An Ecomodernist Manifesto*. Published in 2015, the manifesto is coauthored by a number of writers whom we've already mentioned (Shellenberger, Nordhaus, Lynas, Ellis), as well as others whose hobbies are worth learning about, such as Stewart Brand (author of an "ecopragmatist manifesto" from 2009 wherein he defends the unregulated engineering of ecosystems)[20] or David Keith (an advocate of solar geoengineering).[21] The keyword of "An Ecomodernist Manifesto" is "decoupling": Thanks to (mainly nuclear) technologies, we would be able to "decouple" economic development from any negative environmental impact—in other words, we would be able to create some sort of technological *off-planet* bubble, of a synthetic world (the idea of a "fully synthetic world" is evoked on page twenty-five of the manifesto) that would completely do away with any dependence on the Earth and—QED—any negative environmental impact. Is this the final avatar of the rallying cry of the ecomodernists to the geo-constructivist discourse and its way of recognizing its place within the Anthropocene with one hand while completely rejecting it with the other? The internal contradiction of the ecomodernists, torn between, on the one hand, the idea that everything is interconnected and, on the other, the desire for terraforming the planet from the outside? A careful examination of Latour's thinking on the subject can help us to respond to these questions.

CHAPTER 7

THE "POLITICAL ECOLOGY" OF BRUNO LATOUR

No Environments, No Limits, No Monsters (Not Even Fear)

Associated with the approach known as actor-network-theory, Bruno Latour's expanded sociology describes the way in which humans and nonhumans are assembled.[1] Rejecting the ontological, social, and political primacy of human subjects, Latour maintains that the power for acting is "distributed" within the networks that the "actants" perform. A politics conscious of the generalized hybridization of things and people should, as a result, promote an "object-oriented democracy."[2] From this rather brief presentation, we can already extract the following idea: There can be no doubt that Latour's thought allowed for questioning, with great efficiency, the way in which we represent—socially, politically, and aesthetically—human worlds. And there are numerous contemporary thinkers that, by insisting on the necessity—counter to a persistent anthropocentrism—of pointing out and saving objects, things, and matter from the hidden backdrops to which we have assigned them, pay tribute in their own way to Latour's theoretical contribution. Nevertheless, in this chapter, we are going to show that Latour's thought generates a certain number of problems. Contrary to what the sociologist of the composition of worlds maintains, we make the claim that "critical" thinking is not a thing of the past: It is what allows us to reflect on how not to form compromises with disastrous realities. If critical thinking must abstain from believing in the transcendence of a so-called real world beyond the world it analyzes, a real world that would be the source of truth that this critical thinking would brandish against the false and obscure appearances of this world, false appearances it would be critical theory's goal to annihilate, then critical thinking must just as resolutely—in contrast to Latour—refuse to believe that immanence is a benediction.[3] Well-honed and duly sharpened, critical thinking cuts through immanence in order to open up other paths for theory and political action.

In order to properly take into account the problems Latour's thought creates for us, we should begin by examining "Love Your Monsters: Why We Must Care for Our Technologies As We Do Our Children," Latour's article from Nordhaus and Shellenberger's anthology that we mentioned earlier.[4] One might ask, why take this essay as our point of departure? Because the piece blatantly offers a condensed version of Latour's thought and, at the same time, demonstrates how his thought works so well with ecomodernism. Latour's article begins with a rereading of Mary Shelley's *Frankenstein*, a hermeneutic operation that had already been done ten years earlier in *Aramis, or the Love of Technology* (1992). In that book as well as in his article, Latour affirms that Dr. Frankenstein's crime is not his hubris, his transgressive creation, but rather his abandonment of his creation. His creation became monstrous because of being abandoned, and not as a result of its atypical genesis. *Frankenstein* will therefore be paradigmatic of our deficient relationship with what we create: By naming certain foods *frankenfood* or *frankenfish*, Latour tells us, we continuously reproduce our contempt for Dr. Frankenstein. Instead of caring for our creations, we reject them. And it's precisely this rejection that is, in the end, the cause of our misfortunes.[5]

This rejection will be anchored within the way we consider ourselves as moderns. In the article, continuing analyses that were first taken up in *We Have Never Been Modern*, Latour claims the perception we have of ourselves rests on a misunderstanding: We believe that we have been capable of clearly maintaining a separation between the areas of science and politics, and we believe that modernity consists precisely of this separation; but in all actuality, we have done nothing but spend our time constituting our world on the basis of endless hybridizations.[6] In reality, so-called modern science would have entangled science and politics, humans and nonhumans, and this modern science would have created and churned out what Latour calls "attachments": connections between nature and scientific production (GMOs), links that would have never been made possible without robotics (Latour takes as an example robots sent to Mars), climate change, etc. To think of oneself as modern would then be to think that science would have "emancipated" us from "nature," when really science only created all the more imbroglios and entangled messes for us to deal with. In this light, Mary Shelley's novel would absolutely and positively be a prisoner of the belief that modernity created of itself. And prisoners we would be, as we once were and still are always and forever: And even though Latour tells us that there is a dissonance, shining bright in the light of day, between what we believe ourselves to be and what we truly are, and even though everything turns out to be interconnected and attachments become more and more evident, we refuse the latter. We refuse this generous ecology that attaches humans to nonhumans—to GMOs, to bacteria, to the Earth. Why?

Because modern belief, as erroneous as it may have been, still had its consequences: Believing that science had emancipated us from nature, we have believed in the existence of a Great Divide between us and the rest of the world. There was what we were making, our technologies, progress, and, alas, all the collateral damage of this progress—pollution, the hole in the ozone layer, Chernobyl. But the idea itself of collateral damage would be a result of this Great Divide. Furthermore, Latour maintains that ecological disasters are analogous with Frankenstein's monster in that our representations also rest on a Great Divide: culture-technology-humans *versus* nature-nonhumans. However, everything changes, or everything would change if we could allow ourselves to see that *everything is attached.* Everything would change if we could understand that science has not stopped connecting nature and culture, to the point of creating what Latour calls "nature-cultures."[7] If, in fact, everything is connected, then collateral damage or, at the very least, "the unwanted consequences" of progress are inevitable: Whether we like it or not, any new technology will inevitably have an impact on us as well as the aforementioned "environment."

As with many thinkers who came before him (Michel Serres, Ulrich Beck), Bruno Latour calls for the expiration of the concept of the environment: This concept is built upon a fallacious divide—but one that has real consequences—between humans (at the center) and their environment (surrounding them).[8] In other words, as Latour claims, we began to speak about the environment at the moment when we came to realize that there is no such thing as the environment! And yet, with the conceptual and real disappearance of the environment, what also disappears, as if by a spell, are the monsters: *The monstrous proves itself to be normal.* Instead of being surprised and distressed toward the "unwanted consequences" of new technology thrown into the global marketplace, we should, on the contrary, take up the utmost and constant care regarding what we create, following and tracking the consequences of our creations. Far from simply abstaining from creating, far from "let[ting] the humans *retreat*—as the English did on the beaches of Dunkirk in the 1940s,"[9] and leaving so-called nature all alone in its own corner of the world, Latour challenges us "to intervene even more." In this sense, "the environment is exactly what should be even more managed, taken up, cared for, stewarded, in brief, integrated and internalized in the very fabric of the polity."[10] We must become even more the "master and possessor of nature" to use the famous phrase from Descartes, if we understand that this "mastery" must be considered as an "attachment" that is becoming more and more close-knit between "things and people." More attachments, more mastery, more interventions—such is the agreement—as we have seen—someone like Paul Crutzen would like to put in place.

The "political ecology" being sketched out here seems as far away as possible from that of someone like Hans Jonas: Where Jonas, in his *The Imperative of Responsibility*, lashes out at a "definitively unbound Prometheus"[11] and, along with the majority of environmentalists and ecologists, calls for concerning ourselves with limits, Latour, in complete contrast, develops the idea of a guilt-free, uninhibited Prometheanism, a kind of super-Prometheanism—in other words, a hypermodern version of modernity. Here, we moderns who consider ourselves as modern, as well as all the Hans Jonas types, and everyone else who wants to "retreat within themselves" as fearful religious spirits, we're supposed to state: *"Thou shall not transgress."* Latour sings the praises of his friends Nordhaus and Shellenberger, who have—according to him—the courage to affirm: "We shall overcome."[12] So we must rise above our timid ways of technologically intervening, rise above the consistent fear of being the Prometheus that we can't help but be, we who are technologically assisted, terraformers of our own planet. It should now be crystal clear to us what ties Latour's thought to the post-environmentalism of Ted Nordhaus and Michael Shellenberger: Showing that everything is connected is the best way for affirming the idea that the entirety of nature has been anthropized. Consequently, since, in recalling the claims of Nordhaus and Shellenberger, there is no environment that we can consider as a "thing" separate from humans, why should we even speak about environmentalism? Let's be post-environmentalists and affirm that "political ecology" is not afraid of monsters *since monsters don't exist* or are simply an effect of the lack of attention to what we are doing.

RESOLUTELY MODERN

One point, however, clearly appears to distinguish Latour from Nordhaus and Shellenberger: Where the former criticizes the false consciousness of the first modernity, this modernity that is not "reflexive" and that has not integrated the inevitable "risks" contained within it, the latter two simply swear by a unilateral and immediate "theology of modernization." Nevertheless, Latour, the ecomodernists, and the post-environmentalists all share the same target: those who refuse development; in other words, those who would like to move "from *hubris* to asceticism," Latour writes, in the longer version of "Love Your Monsters" that also bears the title "It's Development, Stupid! Or How to Modernize Modernization."[13] In one of the footnotes, Latour specifies, for North American readers, "De-growth is the term used by certain French groups" for describing asceticism. It's clear that for Latour, those people who would like to "retreat" or defer from development resolutely find themselves definitively snared by modernist belief, by the way in which they also think that there is a Great Divide between humans and nonhumans. In *Politics of Nature* from 1999 [2004], ecologists and environmentalists were attacked for their "nature-centrism" supporting an unchanging modern belief in a nature that we must conserve intact. An extreme example of the figure Latour refers to is the deep ecologist: "Deep ecology, in my interpretation, is situated as far as possible from political ecology."[14] But in 2011, Latour sets his sights on attacking the idea of de-growth. De-growthers are considered too "stupid" to understand the necessity of development. Of course, for Latour, everyone who believes in the Great Divide between humans and nonhumans is stupid; but the ones who are really idiotic and dangerous are those who believe in this Great Divide and also confirm it, and reinforce it, because of their very asceticism. They are dangerous in that they don't understand to what extent, in order to deal with environmental problems, we must intervene even more, technologically speaking. Just like Nordhaus and Shellenberger, Latour maintains that we must do away with "the limits of the notion of limits," and that the time has come to "develop *more*, not *less*": "The goal of political ecology," Latour writes, "must not be to stop innovating, inventing, creating, and intervening."[15] Like Nordhaus, Shellenberger, Ellis, and Lynas, Latour firmly believes that our salvation goes by way of technological development. In this sense, and this is precisely the point we must grasp here, *Latour is resolutely modern*—without any distance from it.

Of course, Latour claims that he is not modern and that "between ecologizing and modernizing, we have to choose."[16] In examining "An Ecomodernist Manifesto," Latour will finally indicate his position: Even if he admits to "having always been a postenvironmentalist," it is impossible to make an apology for "decoupling," since everything is interconnected—decoupling is a pure "anachronism" in the age of the Anthropocene.[17] But let's apply to Latour (even if it's not very nice) the same theory he applies to the moderns (they believe in separation, but they are creating hybrids) as well as the ecologists and environmentalists (they consider themselves to be speaking in the name of nature, but that's not really the case): Latour considers himself as not being modern, but his theory demonstrates the exact opposite. In fact, by maintaining that science or, more specifically, the development of technoscience will save us from environmental problems, Latour expresses the essence of modernity to the point of a hypermodernity. We should, in fact, note that, in his analysis of "An Ecomodernist Manifesto," Latour doesn't articulate *any critique whatsoever of technologies*: He asks for the ecomoderns to shed some light on who they have as "friends" and "enemies," which is rather shocking since this light has for a long time now been shown by Nordhaus and Shellenberger. And all it takes is to simply glance at any of the pages on the Breakthrough Institute's website to know that the friends of the ecomoderns are those who believe in GMOs, nuclear energy, synthetic biology, and the economic model that underpins this relation to technologies, and that their enemies are environmentalists, de-growthers, Naomi Klein, etc.[18] Would Latour ask of his friends to name their friends and enemies that he himself appears so hesitant to do?

THE VON NEUMANN SYNDROME

Latour thinks that he has turned his back on modernity, but in reality he simply confirms what von Neumann perfectly summarized in 1955:

> Finally, and, I believe most importantly, that the prohibition of technology (invention and development which are hardly separable from underlying scientific inquiry), is contrary to the whole ethos of the industrial age. . . . It's hard to imagine such a restraint successfully imposed in our civilization. Only those disasters that we fear had already occurred, only if humanity were already completely disillusioned about technological civilization, could such a step be taken. But not even the disasters of the recent wars have produced that degree of disillusionment, as is proved by the phenomenal resiliency with which the industrial way of life recovered—or particularly—in the worst-hit areas.[19]

The example of Fukushima perfectly validates this remark. The questioning or challenging of technology is the prohibition in front of which all moderns recoil, Latour included—and this is what we can call the von Neumann Syndrome.[20] In the words of Günther Anders, this syndrome can also be described in the following manner:

> What can be done must be done. . . . The possible is generally accepted as compulsory and what can be done as what must be done. Today's moral imperatives arise from technology. . . . Not only is it the case that no weapon that has been invented, has not also effectively been produced, but every weapon that has been produced, has been effectively used. *Not only is it a rule that what can be done, must be done, but also that what must be done is inevitable.*[21]

In this sense, "It's development, stupid!" is the modern slogan par excellence: the statement that prohibits preventing the passage from invention to implementation and from implementation to usage. *Moderns are those who believe in the sacrosanct saintliness of an inevitable technological development.* But this supposed inevitable characteristic of development hides a specific trait of modern technology: its relation with the impossible. What it becomes a question of developing is the capacity to technologically overcome the impossible, and the modern belief that consists of considering as possible the overcoming of the impossible. Latour wholeheartedly agrees: The moderns "want the impossible, and they are right in wanting it"; but he also adds, we must change the impossible to which we're referring.[22] According to Latour, we must abandon the impossible vis-à-vis nature; and this shouldn't take much effort to do since, for Latour, this emancipation—this detachment—*never took place*! In fact, Latour maintains that the moderns, without being aware of it, passed their time hybridizing humans and nonhumans. All of a sudden, emancipation in relation to nature has become impossible; however, the impossible that we must promote, according to Latour, is the one consisting of doing away with "the limits on limits": Wanting there to be no more limits, this is the impossible that Latour calls forth from his desires. But to "do away with the limits on limits," is this not in fact the very essence of the modernist project? Do we not already find the origins of the von Neumann Syndrome in the work of Francis Bacon? In his *New Atlantis*, Bacon writes:

> The end of our foundation of knowledge of causes, and secret motion of things; and the enlarging of the bounds of human empire, to the effecting of all things possible.[23]

Contrary to what Latour claims, for the thinkers of modern science, the problem is not so much an emancipation vis-à-vis nature—this has already theoretically been accomplished by the transformation of nature into a mathematical object, into abstract quantities (i.e., detached from sensible material)—but rather the future technological creation of all imaginable inventions. It's this very impossible that Latour, as a good modern, has not detached himself from. It's this very *attachment* that leads him to make the remarks at the conclusion of his article: "We want to develop, not withdrawal."[24] So here we have what places us in a position of an infernal dilemma: Either we have the proposition of a developmentalist ecology and environmentalism *without restraint* posed by Latour and his ecomodernist and post-environmentalist comrades, or we have some form of withdrawal, some kind of complete refusal of *any or all* technology.

This alternative to which we feel ourselves to be constrained should be a bit surprising since Latour appears as some sort of sage who knows perfectly well that progress is indistinguishable from "unwanted consequences" inherent with the turbulence of the world—but what consequences does he draw from this necessity consisting of heeding to the consequences? What does Latour mean when he says take care of our creations *throughout their consequences*? Describing the precautionary principle as it is introduced into the French Constitution (the Barnier law [loi Barnier] concerning the protection of the environment), Latour strives to show that his adversaries, as much as those who support him, are already pretty much in agreement. The first group complain that this law, the precautionary principle, will lead to anticipating so many potential risks that it will end up preventing innovation, whereas the second group—those whom Latour, so as to distance himself from them, calls "modernist environmentalists"—congratulate themselves for creating a law dictating "no action, no new technology, no intervention unless it could be proven with certainty that no harm would result."[25] And yet, Latour refuses to give in neither to paralyzing anticipations nor to any sort of a "withdrawal" but instead demands for—what exactly?—*a continual practice of care-taking.*[26] Hence the metaphor of love for one's monsters: Loving our technological creations consists of following alongside them with their effects, not simply letting them grow all alone by themselves in a corner, of not taking into consideration that they will have unexpected consequences; consequences that they most certainly will have, because our creations are attached to us, to everything, to the world, because of the very fact of a generalized interconnection that links each part of the world with every other part of the world.

Naturally, the idea of not letting a nuclear power plant run all by itself and simply shutting the doors behind oneself before heading out to the pub seems rather obvious, as does installing automatic alarm systems capable of encountering "unexpected" consequences. But Latour seems to forget that the precautionary principle has as one of its purposes to install constraint systems that already integrate uncertainty into them! Stating that we have no idea what the consequences will be when we introduce a new technology into the world is completely true, but this is precisely the central issue in our relations with technology: This should elicit what Hans Jonas would call some sort of *concern and responsibility* (Sorge). And yet, this concern consists precisely of thinking *in advance* that a certain uncertainty must lead us to *not* foster certain technologies. Excusing ourselves *in advance*, from such an anticipation, amounts to nothing less than repealing the precautionary principle and venturing out into the unknown, like a good modern. If there's something to reproach in the precautionary principle, it's not to have paralyzed action but to have (wrongfully) thought that risks can be completely calculated (i.e., in the form of the risks of the risks). Against this *precaution*—following the work of Jean-Pierre Dupuy—we must propose *prevention*—i.e., a preventive action that would accept *our inability for calculating risks*.[27] Namely, the problem is not knowing whether or not there will be unwanted risks (that's already obvious) but knowing that we can't allow for certain unwanted consequences at all costs.

And yet, Latour's theory won't allow us to not want. To use Günther Anders's terms, Latour's theory won't allow us to not go from conception to production and from production to usage. In wanting to technologically overcome the impossible, it generates a new impossible: the ability to say no. From now on, for those who adhere to Latour's ecological constructivism, to ecomodernism, to post-environmentalism, to post-preservationism, as well as the ecology of resilience, all that remains is the possibility of stating that each action within the world will have unexpected consequences, be them good or bad, that will then require new actions. Within this infernal loop, Prometheus, whose name means "fore-thinker," fuses with his brother, Epimetheus, "the one who is the after-thinker." In *Technics and Time*, volume 1, *The Fault of Epimetheus*, Bernard Stiegler insists on the difference between Epimetheus, who, since he can't foresee anything, leaves human beings with a deficiency, without qualities, and Prometheus, who, because he is the one with foresight, takes care of human beings by compensating for the fault of Epimetheus (the fault of having forgotten to provide the human being with natural qualities) through the gift of technics.[28] With the dangerous fusion of Prometheus and Epimetheus, it is no longer a question of *care* and anticipation: Henceforth, Prometheus can make use of his weapons arsenal and (nuclear) fire, only now his arsenal is precisely unchained from the necessity of foresight. Epiprometheus is this new monster—do we have to love him? Must we attach ourselves to him? Let's hesitate a bit before responding; afterward, it will no doubt be too late.

DR. FRANKENSTEIN HESITATING AT THE THRESHOLD OF MODERN MAGIC

Let us hesitate. And return to the analysis that Latour made of Mary Shelley's *Frankenstein*. It is, of course, true that Victor Frankenstein's creature suffers from a lack of love from his creator. When they both come to encounter each other in the Alps, Victor Frankenstein calls his creature every name in the book, "a devil," "a vile insect," "abhorred monster," etc.[29] And the poor creature shares with him his solitude, reproaching his creator for not taking up "the duties that encumber him" toward his creation.[30] We will thus grant to Latour the remark according to which Victor Frankenstein's attitude lacks the most elementary care and appears irresponsible. All right—but why, in fact, did Victor Frankenstein end up in this position?

Contrary to what Latour seems to believe, this doctor is not exactly a modern: In his youth, he was influenced by "natural philosophy," Cornelius Agrippa and Paracelsus—in other words, conceptions of science granting a place for alchemy and magic, searching for the philosopher's stone and the elixir for everlasting life. It's from these sciences and philosophies that Victor Frankenstein draws his fundamental desire: to create life—to become capable of "animating lifeless clay,"[31] "banish disease from the human frame and render man invulnerable to any but a violent death!" But Dr. Frankenstein, during his studies, encounters modern science and the way in which it substitutes itself for magic. And this is what he states to Professor Waldman in Mary Shelley's novel: "The ancient teachers of this science promised impossibilities," whereas "the modern masters promise very little; they know that metals cannot be transmuted, and that the elixir of life is a chimera. . . . These philosophers, whose hands seem only made to dabble in dirt and their eyes to pour over the microscope or the crucible."[32] And in doing this, seemed to have accomplished "miracles":

> They penetrate into the recesses of nature, and show how she works in her hiding places. They ascend into the heavens; they have discovered how the blood circulates, and the nature of the air we breathe. They have acquired new and almost unlimited powers; they can command the thunders of heaven, mimic the earthquake, and even mock the invisible world with its shadows.[33]

A humility of the means and changes in the methods used compared to the alchemists, no doubt, but all this is done so as to achieve all the more wondrous results! In Mary Shelley's novel, modern science manifests itself merely as the pursuit of magical desires by other methods. And us as well, how far have we really come from exiting this search for an elixir of life? In a famous passage from his *Discourse on Method*, Descartes claims that modern science, more specifically physics, will allow us to attain this "first good," which is "the maintenance of health."[34] And today, rather than health (which implies a state for one to attain) we speak more about fitness, which in and of itself knows no absolute limits: As Zygmunt Bauman writes, "However fit your body is—you could make it fitter."[35] We will, of course, not conflate the elixir of life with the maintenance of health and fitness; nevertheless, a certain imaginary of eternal life found a new skin via the flourishing of nineteenth-century science. As Mumford maintains, "magic was the bridge that united fantasy with technology";[36] and science has on several occasions borrowed this bridge to go in the opposite direction—from technology to a dream suddenly become probable and palpable. This hypothesis can help us to understand that what drives us to believe in the almighty power of development and technologies is a belief in modern science underpinned by a desire for magic [vœu magique].[37] Is this belief and this desire not precisely what the ecomoderns, Latour included, perpetuate along with the post-environmentalists? If they are modern, then it's clearly not simply because they are for the creation of even more technologies (that's simply a symptom) but because they don't question in the slightest the way in which modernity has metabolized—integrated, reprised, and modified—the research of magic and its libidinal investment in machines. It's this belief that also animates the wish that consists of being able to control climate change thanks to the technological "optimization" of the climate. And it is this belief that ecological and environmental thought must rid itself of today.

All of sudden, it becomes rather problematic to generalize Latour's proposal consisting of loving *all* our monsters. Indeed, Victor Frankenstein somewhat resembles our monstrous Epiprometheus, who first burns down all of our forests and then reflects on it afterward. We understand quite well that there are cases when, once a creation has been made, it becomes necessary to then take care of it. For example, here we could think of the fictional example (that is becoming less and less absurd as a possibility) of human clones: Once the clones are generated, and even in spite of the law against this already put in place, it would be untenable to not do our best to take care of them. But what Nordhaus, Shellenberger, Brand, Ellis, and Latour's proposition makes impossible is *preventive action*—i.e., the possibility for *not realizing* a technology. We indicated earlier that a fundamental prohibition weighs on our shoulders: the questioning or challenging of technology. To call into question must signify, theoretically, to *return to its cause*, to question the causes, and therefore to also take some distance from the "pragmatic ecology" proposed by someone like Émilie Hache, since this pragmatic ecology is propped up with a "pragmatic philosophy" that is an "art of consequences interested in the effects these propositions induce."[38] The ecological thought that we are seeking to promote is antipragmatic for motives that are . . . very pragmatic: Being interested in the effects is to be interested too late; we must reflect and consider the causes prior to our actions. Only a return to causes, to the ends, to the principles, and to what we desire can allow us to make a distinction between technologies we want and those we don't want. We must make this distinction before, and not after, the fact.

In order to make this distinction, we must think technologies by way of their singularities. In other words, we must stop confusing—as Latour does—the education of children and the production of technologies, the love attributed to the former and the attachments that we can sometimes attribute to the latter. A robot is not exactly the same as a child! Then, we will be able to take a step back and abstain from perpetually producing technologies. It's this act of abstaining that is impossible for Latour, for whom the precautionary principle "is not a principle of abstention . . . but a change in the way *any action is considered*"—in other words, a change in the relation between science and politics. Thanks to this principle, "unexpected consequences are *attached* to their initiators and have to be followed through all the way."[39] The post-environmentalist's problem is that he is always arriving after the party is over: New technologies are released out into the market, and the post-environmentalist begs for us to love them, that we keep close watch over them, and through all the way. In only focusing one's attention on the consequences, any new technologies are given carte blanche (GMOs, fracking, nuclear energy, and climate engineering) so as to leave the door open to all sorts of scientific "controversies." Subsequently, how is any sort of political ecology still possible?

UNCERTAIN COLLECTIVES AND THE POWER OF THE TWO

The incapacity to, truly and properly, carefully assess all the various new technologies is to understand one of the symptoms of Latour's constructivist "political ecology": His unilateral taste for association, composition, and attachment makes access to any dimension of separation, division, or opposition rather difficult—in spite of certain very recent declarations made by Latour.[40] We should remember that for Latour, human beings do not form a society of subjects cut off from the rest of the world of objects but are composed of an evolving collective of humans that recognize the existence of nonhumans claiming a place within this collective. The collective is a set of procedures whose function is to "*represent* the associations of humans and nonhumans through an explicit procedure, in order to decide what collects them and what unifies them in one future common world."[41] We can always still question the collection, accepting or refusing new entrants, inscribing a "jury for each *candidate* of common existence" after "consultation" of these new "propositions."[42] The collection is endless; we only create temporary limits. We should add that this collection is not the result of some sovereign decisions made by humans: Nothing could be more false than the "concept of a *human* actor *fully in command*."[43] Human creators and constructors must accept that "they share their agency with a crowd of actants over which they have no control nor mastery": substituting for, and in the place of, the fantasy of human mastery reigns "uncertainty," which forms the ontological core of the *actor-network-theory*.[44] This uncertainty precedes all construction, all "collective" composition, knowing that the goal of the "political process" is to create a common world: "The unified world is a thing of the future, not of the past. In the meantime we all are in what James calls the 'pluriverse.'"⁴⁵ The goal of this common world is for there to be a guarantee whereby "humans and non-humans are engaged in a history that should render their separation impossible."⁴⁶

It's one thing to have to be concerned with nonhumans; but it's an entirely different thing for any form of separation to be impossible. To propose the commons as *political* objective, nonseparation and unification, is indicative of a crucial point: Latour's *ultimate* objective is not about letting the multiplicity of actants be but about *producing the One*. Certainly, a unified world is a "thing of the future," but such an idea forces us to reconsider the theme Latour holds to in his *Politics of Nature*: "*The* collective signifies 'everything but not two separated.'"⁴⁷ From this, we can draw forth the following lesson: The exclusion of the Two inevitably leads to the One, however in the "future" it may be. Sure, the One is differed since Latour's sense of the collective is uncertain, always "in the midst of expansion," capable of endlessly being amended, reformed, *to be remade again [à refaire]*, always capable of recognizing the existence of newcomers. But once the Two and separation no longer have a place, everything tends to become mixed together and becomes indistinct: *The Latourian collective places all beings on the same plane*—prions, primates and humans, even, as we have seen, technological production is placed on the same level as human generation. Certain people will no doubt find all of this rather silly and foreshadowing the rapturous object-oriented ontologies and their endless lists.⁴⁸ But the problem is that the condition of such a collective makes it impossible to politically discern between what is important and what isn't, placing on the same level the recognition or disappearance of prions with the recognition or disappearance of primates. As Alain Caillé, in his article about Latour, writes:

> It's difficult for us to see which people are susceptible or, more generally speaking, which human subject will be susceptible for a longer time period by the fact of being placed on the same plane as any given electron, amoeba, virus, or a monkey wrench. The lone moral that seems to subsist through the Latourian (de)construction is that of the permanent opening onto the infinity of the thinkable and the doable. Let's translate: everything that can be made, technically, must be made. Nothing will be nor should be able to oppose the indefinite extension of biotechnologies. . . . It's against this absence of limits, this hubris, that environmentalists fight. . . . Is there not something paradoxical in aligning oneself with political ecology in order to *in fine* accredit, at a startlingly rate, objectives that seem completely at the antipodes of the true environmentalists that one encounters in the flesh?⁴⁹

"Nothing in Latour's discourse," Caillé adds, "allows for any kind of resistance to GMOs, to the hegemony of biotechnologies, or, for example, to the indefinite modification of the genome," but everything encourages realizing these technologies.[50] This is for a very simple reason: For Latour, politics never signifies *conflict* (there would have to be, at least, two!) but always a *process* and production—the way whereby the multiple converges toward the One. His problem is that of knowing "how to bring the sciences into Democracy"—this is also the subtitle of his book *Politics of Nature*—and not how to bring Democracy into the sciences. So it's about building, and building better—"How can it [the world] be built *better*?"[51]—but certainly not opposing what could eventually be an ill-conceived creation, that, in hindsight, it would be a bad idea to build. That would mean being opposed to the von Neumann Syndrome: the imperative of absolute realization, this very modern quasi-Baconian superego consisting of bringing into existence everything possible of inevitably being achieved technologically. Latour's collectives without limits are the result of this imperative: Everything is uncertain, except for the certainty of the production of everything.

Consequently, we can begin to understand the problem: Uncertainty, which does indeed adequately describe an ontological state of matter in its quantum state, is utilized in an expansive manner in order to justify the fact that the way in which humans associate with nonhumans is uncontrollable. And yet, from where does the uncertainty arise in the case of Fukushima? From the incredible *agency* of plankton and shrimp? From some sort of mystical and improbable communion between these sea creatures with the just-as-mysterious components of the nuclear reactor? No. A nuclear power plant installed on a seismic fault line, falsified inspection documents by TEPCO, a lack of respect for the international norms relative to nuclear energy and environmental security, the destruction of the cliff that constituted a natural protective barrier, and so on and so forth:[52] Fukushima wasn't an accident, an unexpected consequence linked with uncertainty and to the absence of our ontological mastery over the turbulences of the world; Fukushima was a programmed accident. We knew it, or we could have known. To speak here, in this case, of uncertainty, of some sort of "surprise" or "unexpected" consequences, is either a question of ignorance or the indirect justification of disasters and their causes. Behind Latour's uncertain collectives stands the certainty of pluriverses aspiring for the One; counter to this, an ecology of separation must be capable of reintroducing the power of the Two:

1. The power of separation without which every association becomes indistinguishable. And yet, it's always humans that benefit from this lack of distinguishability, as if it were impossible for nonhumans to exist without humans

and to create links between themselves, nonhumans with nonhumans; as if humans, as Nigel Clark notes, found themselves incapable of recognizing an ontological power that exists outside of them and in relation to which they should recognize their own dependence;[53]

2. The power of opposition without which production and development lose all of their meaning, pure processes without an ethical end. Politically, the power of Two requires not only collectives or associations: it requires actions.

CHAPTER 8

ANATURALISM AND ITS GHOSTS

THE HOLY ALLIANCE AGAINST THE IDEOLOGY OF NATURE

How is it that a "political ecology" like Latour's can lead to the advocation for technological monsters? To post-environmentalists assuring us that there is no such thing as the environment and that we must soon enter into an era of post-preservation? To ecomodernists hedging their bets on the possibility of a "super-Anthropocene" where we will soon have the ability for "decoupling human development from environmental impacts"?[1] To eco-pragmatists assuring us that there is no such thing as a "common world" and it's therefore up to us to build it? One and the same affirmation: Nature is dead. "Thank God," Bruno Latour tells us.

> Thank God, nature is going to die. Yes, the great Pan is dead. After the death of God and the death of Man, nature too, had to give up the ghost.[2]

We should be careful to note that what's important for the ecomodernists is not assuring us that there is *no* such thing as nature but that there is no nature *any longer*. There will no longer be anything such as nature because from now on it will be impossible to encounter a part of nature that hasn't already been touched, utilized, modified, hybridized—by humanity. In the end, this is what the Anthropocene would truly signify: the death of nature by ingestion inside the biotechnological belly of *anthropos*.

However, we would be wrong to think that this idea originated with the ecomodernists. In *The German Ideology*, Marx and Engels already urged us to see the "sensible" world not as "an object directly given for all of eternity" but rather as "the product of industry and the state of society": "The nature that preceded human history is not the nature in which Feuerbach lives, it is nature today which no longer exists (except perhaps on a few Australian coral islands of recent origin)," the authors ironically concede.[3] Here we will begin to see the sketched-out contours of a strange holy alliance between original Marxism (the one preceding the investigations in *Capital*) and ecomodernism on the subject of nature: Nature would be an illusion, an ideological object for liquidation—in the Anthropocene, even Australian coral islands will have become industrial products.

1. *Nature-transcendence.* An example of this strange holy alliance can be seen at work in what Timothy Morton, an author who brings together ecological studies with work in object-oriented philosophy, defends by what he calls a "an ecology without nature." For Morton, the ideological illusion that leads to believing in the existence of nature can be directly linked back to Romanticism, which put nature "on a pedestal": In fact, it's this idea of nature as a "fantasy," as a "metaphor" for everything, as "a transcendental term in a material mask," of which it becomes necessary to rid ourselves.[4] In the end, the Romantic conception of nature as an "ideological construction" of a transcendent outside (much more so than simply being transcendental) will be an accomplice of capitalism:

> And if, finally, Nature as such, the idea of a radical outside to the social system, was a capitalist fantasy, even precisely, *the* capitalist fantasy?[5]

2. *Nature-harmony.* In the same vein, Slavoj Žižek—an author who conjugates Marxism, Lacanianism, German Idealism, and a hefty amount of verbal juggling—sees ecology or environmentalism as the "new opium of the masses replacing declining religion," susceptible to quickly becoming "the predominant form of global capitalist ideology."[6] Harbinger of "ideological mystifications," ecology would also give birth to New Age mysticism, to a neocolonialism of hyperdeveloped Western nations who continually require other emerging nations to limit their rate of development, or to strange liberal communists who buy green, environmentally friendly products but in no way hinder the mode of capitalist production. Žižek tells us that the problem is that ecologists *believe* in a nature as a balanced order, as a beautiful homeostatic totality. However, this sort of nature—this ideological nature—doesn't exist: The lone existing "in itself" is not of the natural order but rather a "meaningless composite of multiples" that should lead us toward "*an ecology without nature*," even considering that "the ultimate obstacle to protecting nature is the very notion of nature we rely on."[7]

3. *Nature-hoax.* We should note that these sorts of remarks would be exactly what certain non-Marxist and clearly antiecologist or anti-environmentalist minds would like to hear: In a 2004 novel called *State of Fear*, Michael Crichton—the well-known author of *Jurassic Park*—portrays environmentalists either as something like Hollywood stars detached from reality having no anthropological

knowledge (a seemingly Rousseauian actor will end up being eaten by a tribe of cannibals), or as greedy members of NGOs *who don't believe* in ecology, or as ecoterrorists, who in reality *know* that there is no such thing as any danger of climate change and thereby attempt to *make people believe* there is such a danger by way of producing artificial disasters.[8]

To sum it up: For Crichton, as well as for Žižek and Latour, it's about making us believe that ecologists and environmentalists believe that nature is a transcendent entity, some sort of beautiful harmonious totality, or that there is within our concept of nature a kind of deceptive hoax masking various vile political interests. However, such a belief in some sort of harmonious totality is much more the exception than the norm. There's not a trace of this nature-harmony or nature-transcendence in the work of the Intergovernmental Panel on Climate Change, who are attempting to capture the future of atmospheric chaos, nor can one find it in the work of the journalist Naomi Klein mapping out the political struggles throughout the world fighting against extraction and promoting regeneration, or in the work of Gary Snyder, poet and radical environmentalist as well as a promoter of bioregionalism. Reinhabiting abandoned territories, Snyder tells us, does not consist of returning to some sort of primitive indigenous condition but rather consists of inventing a "deep environmental ethics, a strong connection with the site, an engagement with the community." Snyder reminds us that the oldest ecologists or environmentalists have always understood the "impermanence of the universe."[9] Since a "deep" ethics is at stake, let's not hesitate from citing Arne Næss, the philosopher who invented the term *deep ecology* and who invites us to "strive for a greater familiarity with an understanding closer to that of Heraclitus" where "everything flows," "abandon, fixed solid points."[10] This rapid overview—which will be further developed in Part III of our present book—seeks to clearly show that the paradigm of a nature that is smooth, organized, and well balanced has been surpassed by the *thought of ecology itself*. From the early founding texts by Holling that we studied, since the "Clementsian Paradigm" and his dream of environmental stability began to be questioned, it's now environmental thinking itself that, beginning in the 1970s, generated an "ecology of chaos" (Worster).[11] The concepts of the ecosystem, or a superorganism, or a climax have all been rejected: There is a movement toward neither maturity nor an end goal in nature. Instead of wanting to show that human beings are the ones devastating nature and disturbing its order, these ecologists of chaos insist on the fact that nature itself is a *natural disturbance*: Fire, wind, predators, and invasive species are all presented as elements of nonhuman disorder. This individualist approach in no way whatsoever recognizes an "emergent" property that would allow us to speak of some kind of a whole greater than the sum of its parts. In regards to the ecologists of the 1980s, Worster writes, "When they look at a forest, the population ecologists see only the trees."[12]

A conclusion becomes clear: The holy alliance of a primitive Marxism devoid of illusions alongside an antienvironmentalist ecomodernism leads, at the level of ideas, to a rearguard battle. The ideology of nature-harmony, of a homeostatic nature, of a nature-substance supposedly remaining unchanged throughout the course of time, has endured; fighting against it can only serve the interests of capitalism for which everything that is stable and substantial is merely an obstacle to tear down. Consequently, what is the deal with this relentless tenacity of the ecomodernists for wanting to kill what is already dead? Is it because the ecomodernists clearly have a fear of zombies? And what if the "death" of nature were not some recent development but a *serial murder* revealing one of the most oppressive tendencies of the West? And what if the fight-against-the-ideology-of-nature was nothing more than an avatar of this tendency?

THE ANATURALIST DRIVE AND THE FOUR DEATHS OF NATURE

Let's dare to pose the following hypothesis: Far from constituting an original or heretical thesis, to envision ecology as against (Žižek) or without (Morton) nature would be a symptom of the *anaturalist drive* of the West, a very specific death drive that would strive to symbolically and truly do away with nature and that would necessarily be translated through various environmental mishaps and disasters; a drive that would endlessly produce its chanted transcription: "Nature no longer exists." How would one go about supporting such a hypothesis?

1. *Logical Tendency.* One could, following the example of Paul Feyerabend, go all the way back to Parmenides for establishing the philosophical origin—or rather the prephilosophical origin—of the death of nature: Abstract, homogenous, distant from any sort of experience, nature, according to Parmenides, already appears as being quite far away from the world of the living. For Feyerabend, this nature-substance would be based on the repression of the experience of the world that was established—for example, in the Homeric world—on the basis of heterogeneous sensations only leading to "aggregates" without a logical unity. We will have to wait for Prigogine's thermodynamics, Feyerabend says, in order for the Western repression to fully give way.[13] For Whitehead, the constitution of a nature-substance would also be at work in Aristotle's philosophy:

> Aristotelian logic has led to an engrained tendency to postulate a substratum for whatever is disclosed in sense-awareness, namely, to look be-

low what we are aware of for the substance in the sense of the "concrete thing."[14]

Indeed, we can credit philosophy's *logical tendency* with the first instance of *physicide*. We can thank Logic for being the weapon thanks to which nature is transformed into an object of science. We should nevertheless emphasize that our argument has no intention of discrediting abstraction; but as Whitehead maintains, we want to simply posit the reminder that philosophy must help us so as not to conflate abstraction with reality—in other words, help us continuously correct our abstractions[15]—in order to leave a place for what we will name, in the third part of our present book, *real nature*.

2. *Monotheistic Substitution.* Monotheisms have provoked a second death of nature in subordinating it to a supreme Creator. In fact, nature remained a model for the paradigm of antiquity, to which Aristotle's famous remarks attest: "And in general art (techné) either completes what nature is incapable of completing or imitates it."[16] But with the monotheisms, nature loses its status so as to become, as the theologian Richard Hooker claims in 1593, the privileged "instrument" through which God can express his commandments, implying from then on "the obedience of creatures unto the law of nature."[17] In this light, when Latour tells us that "Thank God, nature is going to die," without knowing it, he is celebrating a theological killing: the substitution of God for nature as the supreme ontological principle. Already denounced by Schiller in 1788 in a poem entitled "The Gods of Greece," this substitution leads to "stripping away" from nature its "divinity" and, thereby, preparing the terrain for the third act.[18]

3. *General Mechanization.* The third act plays itself out during the seventeenth century: With Bacon, Galileo, and Descartes, nature becomes a mathematizable, quantifiable, measurable, and thus exploitable material in anticipation of the economic progress of human civilization. For Heidegger, we have made nature "available" for our "expectations,"[19] whereas Adorno and Horkheimer see within *Aufklärung* a machine of unification requiring nature as well as society to conform "to the criteria of calculability and utility," to the rules of "abstract quantities."[20] In a similar way, according to the historian Lewis Mumford, it's the physical science of the seventeenth century that is responsible for the "elimination of quantities" and of the "organic," as well as the creation of a veritable "depopulated world."[21] This elimination is painfully laid *bare* in the well-known passage from Descartes's *Meditations* when a piece of wax, once it has come into contact with fire, "becomes a liquid," losing its finer qualities, its odor, and color to, in the end, become the *thing* that Descartes wants to produce: something "extended, flexible, and changeable" that only a careful

FIGURE 3. Louis Ernest Barrias, *Nature Unveiling Herself Before Science* (1899). Polychrome marble, onyx, granite, malachite, lapis lazuli. 200 × 58 × 55.5 cm. Musee d'Orsay, Paris. Photo: René-Gabriel Ojéda. © RMN–Grand Palais / Art Resource, NY.

"inspection of the mind" can take into account.[22] Becoming a liquid, all of the sensible qualities, as if they didn't belong to the piece of wax at all, *as if* reality bifurcated—to use one of Whitehead's terms—between mathematical abstraction, on the one hand, and sense-awareness without substance, on the other.[23] During the process of ontological liquidation, nature is thereby replaced by (truly formless) amorphous matter. So began the era of the mechanization of the world rising up against the cadaver of organic ontologies. Carolyn Merchant has insisted on the way in which organic cosmologies were informed by the representations of nature in terms of the nurturing mother—prodigious nature, an infinitely generous mother, who lacked nothing.[24] It is these beliefs that burst into flames during the seventeenth century. From that time on, as Descartes would say, nature is no longer "some deity, or other sort of imaginary power" but "matter itself"[25]—a formless matter, but also one without force, incapable of escaping our inquisitive gazes in the way in which a Heraclitus, more than twenty-five centuries earlier, deemed it capable of: nature, as this pre-Socratic thinker claims, "loves to hide" (physis kryptesthai philei).[26] De-deified, liquidated, nature is stripped bare, all the way to the bone, as in Louis-Ernest Barrias's bronze sculpture, *Nature Unveiling Herself Before Science,* which dates back to the end of the nineteenth century, analyzed by Carolyn Merchant, where we discover a nature in the form of a woman offering her bare-breasted body, prostituting herself for science.[27]

4. *Global Commodification.* Far from limiting itself to a single culture, a certain part of the world, be it the West or somewhere else, the anaturalist drive is today equally expressed by all societies who participate in the geo-constructivist program: a global capitalist project consisting in exploiting the Earth as if "we" were immunized against any deleterious rapport with the planet. The global geo-capitalist program can be translated as the transformation of nature into commodities (i.e., "dis-embedded" objects, to use one of Karl Polanyi's terms recently echoed by Razmig Keucheyan so as to analyze the financialization of nature). Dis-embedding, writes Keucheyan, requires "(re)constructing reality": In order to change nature into an object that is susceptible of entering into the cycle of the exchange of commodities, one has to isolate within them certain specific parts and to rip them away from their human and environmental context.[28] This is how one can transform water into a service that can be "privatized," or how the pollution of a ton of carbon can be quantified into carbon credits, or how certain risk securities in relation to climate catastrophe can be turned into *cat bonds* (catastrophe bonds), thanks to which insurance companies or reinsurance companies are able to support risks by way of levels.[29] These "real abstractions"—at once cut off (isolated, dis-integrated, in the true sense of the *ab-stract*) from the relational *concreteness* of the world but also *real*, that is, *operational* within precise socioeconomic conditions—constitute a fourth death of nature. It's true, the tendency toward abstraction is as old as logic, and Parmenides, like Aristotle, didn't have to wait for the arrival of financial capitalism in order to forge his categories; but it's as if capitalism simply enlisted the service of logic for its benefit—which is precisely what Marx had already grasped within the following saying: "Logic, is the currency of the mind."[30]

Logical tendency, monotheistic substitution, general mechanization, global commodification: These are the four deaths of nature that lead us to speak of a dominant anaturalism. In this sense, our thesis contradicts head-on the proposition of the anthropologist Philippe Descola, according to which the West would be "naturalist." For Descola, being a naturalist means identifying, under the banner of nature, a part of the world that is completely separate from the rest of it. The anthropologist tells us that there is nothing universal about this position. So, for example, the Achuar don't really leave a reserved place for nature; their cosmology confers most of the attributes of human beings upon animals and plants.[31] This nondualist cosmology would demonstrate the inability of the Achuar to turn nature into an object.[32] From this lone example (Descola provides the reader with a great deal more of them), we will be able to sketch out the general thesis defended by the anthropologist: Only the West represents the nature of nonhuman worlds and the culture of human worlds as incommunicable.[33] Assuming this is true and that it is the West alone that has produced this dualism,[34] it is striking to discern that each time Descola wants to demonstrate an example of Western naturalism, he offers up a radical process of the erasure of nature and not one of a nature-culture divide! He thereby accuses Aristotle's classification method of organisms by composition and division for having produced a "system of Nature in which species are *disconnected* [our emphasis] from their particular habitats and *stripped* [our emphasis] of the symbolic meanings that were attached to them," which Aristotle can only do by way of "*decontextualizing* [our emphasis] the entities of nature" in reducing them to the status of objects.[35] In the same way, "modern naturalism," born in the seventeenth century, will turn nature into something "mute and impersonal."[36] As for Christianity, it turns nature into Creation:

> a provisional scene in a play that will continue after the stage scenery has disappeared, *when nature will exist no more* [our emphasis], and only the principal protagonists will be left: namely, God and human souls, that is to say, human beings in a different form.[37]

"When nature will exist no more": We can't put it any better than that! The denunciations of dualisms and of the great divides are the theoretical tree that conceals the devastated forest. A prisoner of this denunciation, Descola maintains the idea of a humanity faced against nature; however, it's perhaps nature that has lost (its) face: Nature was not constituted as an entity that was integrally *separate* but as an entity that was *denied*, deconstructed and destroyed by the negative waves of repetitive deaths. From now on, what we must not do (which is what Descola does in a symptomatic way) is to see a continuity between Descartes and Darwin,[38] but on the contrary, we must show how Darwin is a figure of minoritarian bodies that were opposed and continue to be opposed to the anaturalist drive. So as to simply develop this example, Darwin will have, therefore, attempted to re-establish what had been erased—the sensibility and intelligence of animals, their "pensivity" as Jean-Christophe Bailly puts it.[39]

PRETEND AS IF NATURE DOESN'T EXIST

Physicide is a widely popular activity for those who consider themselves as belonging to the so-called "West," as well as for a large portion of the entire world. In this light, ecomodernists and post-environmentalists are merely prolonging and intensifying this murderous tradition, whether they find themselves on the side of agreeing with the real abstractions of capitalism or they are spellbound by the wonders of science and modern technologies—or whether, as is very often the case, they agree with both sides. It still remains that the expression "the death of nature" is, at the very least, problematic—whether used by those who celebrate it (Latour) or by those who deplore it (Merchant): Are we not simply dealing with a metaphor? After all, the death of nature is always the death of a representation of nature—but not the death of nature itself! In this sense, there would be no death of nature but simply a change in its representation. We shall study this hypothesis in order to then refute it.

1. Certainly, Merchant shows that it's the organicist representation of nature that has disappeared and that today the task of ecology or environmentalism is to restore this image. In the same way that Bill McKibben shows, in his book *The End of Nature*, the end of this image of nature as "an independent force, something greater than us":

> By the end of nature I do not mean the end of the world. The rain will still fall and the sun shine, though differently than before. When I say "nature" I mean a certain set of human ideas about the world and our place in it. But the death of those ideas begins with concrete changes in the reality around us—changes that scientists can measure and enumerate. More and more frequently, these changes will clash with our perceptions until, finally, our sense of nature as eternal and separate is washed away, and we will see all too clearly what we have done.[40]

So, it would in fact be crazy to think that nature as such, from terrestrial carbon to celestial black holes, would be susceptible to being destroyed by human beings. In other words, from the viewpoint of an ecology during the era of the Anthropocene, the end or death of nature could simply mean the following: modifications and changes made by *human beings* of a *terrestrial nature* change or modify the conception of our *relation* between terrestrial nature and human beings. There is nothing superfluous about this simple statement because the antienvironmentalists desperately cling to this statement, repeating it every chance they get by then saying that what environmentalists claim is nonsense when activists say that nature—the universe, its stars, and its laws—

could be destroyed by human beings. So as to defuse this criticism, we must add that when activists speak of the term *nature* ("as being in danger" and therefore needing "protection"), what they mean by this term is an *expansive synecdoche*: The *whole* (nature in its entirety, the universe) stands for its part (the Earth, the ecosphere, *this specific* lake, *this specific* forest, even *this specific* animal)—the part at play regarding *this specific* political struggle, *this specific* trial against a multinational corporation, etc.

2. Nevertheless, is it enough to merely speak about the death of nature, or rather the deaths of nature, as simple changes in the representation of nature's death? In our opinion, such a hypothesis does not allow for a true understanding of the essence of anaturalism—in other words, the scope of the negative operation that the term *death* paradoxically risks masking. We would do well to remember Crutzen and Schwägerl's essay previously cited in the second chapter of our book:

> The long-held barriers between nature and culture are breaking down. It's no longer us against "Nature." Instead, it's we who decide what nature is and what it will be.[41]

In fact, Crutzen and Schwägerl don't completely deny the idea of nature: Indeed, everything happens as if we still needed nature, if only to deny its existence. But why? Should the anaturalist drive pre-emptively put its foot on the brakes? Should it restrain itself from its own tendency to dis-embed everything? Can the eco-constructivist approach—in spite of itself—help us to respond to these questions? For Latour, what is "dead" is the representation of *nature* as a term that "makes it possible to recapitulate the hierarchy of beings in a single ordered series," it's nature as an "inflexible causality" with "imprescriptible laws."[42] And yet, Latour firmly refuses the idea of a weak constructivism that would look to simply show that "nature does not exist" since it would be nothing more than "a matter of social construction."[43] Latour simultaneously refuses a naturalist realism—therefore a type of substantialism—as well as refusing a sociological idealism that would grant all power, all agency [tout pouvoir] to humans and that would consider nature as a fiction; and yet, in complete contradiction with his best of epistemological intentions, his "political ecology" requires the "creation of space entirely freed from the grasp of nature"![44] In the same vein, Latour maintains that we are within an epoch that is "post-natural, post-human, and post-epistemological"[45]—*but most certainly not post-technological*. This apparent contradiction can be formulated in the following way: We will certainly not dare to immediately affirm that nature doesn't exist (that would mean assuring some sort of clear ontological principle, something that pragmatic ecologists have an overwhelming hatred for), but we will pretend as if nature doesn't exist; we will say something like:

> Far from us to embrace the idea that nature doesn't exist and that it would merely be a social construction; but, in order to found a true nonmodern, pragmatic, and developmentalist political ecology (an association of adjectives that lead scandalmongers to affirm that we are, in spite of ourselves, modern, even hypermodern), in order to propose an ecology capable of taking into consideration the collective of humans and nonhumans that are indeed attached, let's pretend as if nature didn't exist. Don't deconstruct it (that would be too nihilistic), but let's construct a common world as if nothing had happened.

However, as Catherine and Raphael Larrère write:

> The history of "mad cow disease" is an exemplary demonstration of the fact that nature still exists, and that the problem doesn't come from what nature has by default done to us, but from the fact we have pretended *as if* [our emphasis] nature didn't exist, as if there were no longer anything more than machines [*mécaniques*] that existed.[46]

Let's consider this analysis literally and follow it through to its most extreme consequences. We could certainly describe the geo-constructivist and eco-constructivist relationship to nature as something of the order of the "as if," as a sort of *negative hallucination*: We will pretend as if nature didn't exist; we will reduce it to its most raw and amorphous material as if it had no form whatsoever in and of itself; we will simply consider it in terms of a (object) "resource" as if it wasn't also a source as well. But why do this? For ethical reasons? Based on some kind of remorse in the aftermath of the resulting devastations of the anaturalist drive? It seems more easily possible for us to explain this "as if," *this pragmatic scotomization of nature*, starting from two major reasons:

1. The first one is logical and ontological: In order to say that nature is dead or has come to an end, one is still in need of a living interlocutor . . . Which means, moreover, in passing, that it is ontologically impossible to declare the end of nature, since there is no position, no place that exists from which such a declaration could be proclaimed. Until proof to the contrary, until some sort of abiological transhuman or strictly technological singularity reigns without a rapport on a lifeless Earth, the anaturalist drive of the West will be in need of an ecosphere in order to continually revitalize itself while doing its utmost to destroy it. In other words, the anaturalist drive must always be partially thwarted and refuted, since completely denying ecospheric nature would be suicide.

2. The second reason is economic. It's true, the anaturalist drive at work within the geo-capitalist program must be thwarted so there can always remain territories to colonize. After all, if there were truly no longer any nature left at all, there would no longer be anything left to exploit . . . In this sense, capitalism has always needed to concede a minimal degree of existence to nature—under the name, for example, of a natural "resource"—before swallowing it up, exploiting it, or reconstructing or reshaping it. In this sense, pretending as if nature didn't exist could be viewed as the "formation of a compromise," to borrow from Freud: We liquidate nature, but we nevertheless preserve it just enough all the same so we can continue to survive—and overuse or overexploit it.

So, consequently, we should be somewhat suspicious of the expression "death of nature," not because of the fact that it expresses too large of a claim, not because it's an exaggeration (since nature is above and beyond the powers of human beings—they are incapable of modifying black holes), but because of the fact that the expression doesn't allow us to understand the self-limitation of the anaturalist drive: this limitation being *a)* an *operation of symbolic negation* (pretend "as if") *b)* that is achieved through a *technics of isolation*, of decontextualization and of dis-embedment *c)* whose effects are the *partial destruction* of living environments [milieux vivants].

THE WILD ROMANTICISM OF AN UNACCOMPLISHED PAST

Sometimes, the negative hallucinations of the geo-constructivists and the ecomodernists are replaced with positive hallucinations—with ghosts. In fact, everything happens as if nature was a recalcitrant ghost that continuously took on not only new forms, new representations, but also new bodies: We kill it, we proclaim nature's end; nature is reborn, and, in spite of everything, makes itself felt once again as the persistent source behind the "resource" that we strive to reduce it to; we calculate it in terms of "services"; but nature politically makes its return in South America under the name of Mother Earth. Well how about that . . . doesn't that sound like some sort of old-fashioned Romanticism! Is this not precisely the kind of image—of a source, Mother Earth—that a modern (or nonmodern) ecology should strive to get rid of as fast as possible! In 1989, Edgar Morin, wrote:

> The aspiration toward nature not only expresses the myth of a lost natural past; it also expresses the needs *hic et nunc* of persons who feel tormented, oppressed within an artificial and abstract world.[47]

This political romanticism bears no relation to the caricature portrayed by a certain kind of contemporary hegemonic thought. When Carolyn Merchant assigns ecology with the task of showing that, within the system of nature, "each part contributes equal value to the healthy functioning of the whole" and that, as a consequence, "all living things, as integral parts of a viable ecosystem, thus have rights,"[48] she does not imagine, like some kind of geo-constructivist astronaut, a pure nature: Counter to the geo-constructivist death-drive, she affirms a holism of struggle. In contrast to Latour, Merchant is not evoking the death of nature in order to thank God but in order to open up new paths for the living. In the same way, what Edgar Morin calls "the ecosystemic resurrection of the idea of Nature" has no meaning whatsoever if we cut it off from the equation of a project of civilization that Morin wants to base on a "new planetary consciousness"—that is, *a collective consciousness that has never previously existed.*[49] So, if nature, in its terrestrial form, can be resuscitated, it's not by way of some animist or magical process, nor by way of some ontological law that guarantees us the immortality of a nature-substance. We must, on the contrary, re-vive—re-*suscitate*—terrestrial nature. This is one of the tasks of art—think of *Meek's Cutoff* (2010), the film by Kelly Reichardt that revisits the past of the United States in showing what it could have been: ecologically minded, noncolonial, feminist.[50] And this is also one of the tasks of politics—think of McKibben and the 350.org movement, who, in order to fight climate change, ask universities to withdraw their financial investments in enterprises using fossil fuels.[51] So if there is a romanticism worth convoking, it's not one that would prop itself up on an immobile nature-substance but rather a romanticism that desires a world that was promised in the past but that we have knowingly prevented from coming into existence. The ghosts of nature are now holding us responsible for this wild romanticism whose aspirations have been abandoned.

CHAPTER 9

THE TECHNOLOGICAL FERVOR
OF ECO-CONSTRUCTIVISM

ECO-CONSTRUCTIVISM AND GEO-CONSTRUCTIVISM

Now that we have completed our analysis of the anaturalist drive, we can begin to propose a definition of eco-constructivism, so as to include all those types of thought that, arising from out of the paradigm of the turbulence of the world, affirm that:

1. Everything is uncertain in the world, including our actions.

2. Lacking an ability to be the master of causes, we must pay careful attention to the consequences of our actions and the "surprises" that these actions will generate.

3. Causes and consequences form a tightly knit weave of connections that don't allow for the separation of the subjects and objects involved in an action. Attached to a human subject, nonhuman objects also have the capacity to act, they as well have *agency*: They are simply "actants" among other actants, without ontological or political asymmetry, all of them snared within the same network.

4. A post-environmental attitude is the lone reasonable attitude to hold onto within an interconnected world where it is impossible to distinguish between, and separate out, a natural environment as such.

5. In this sense, it is possible to affirm that there is no such thing as nature (as such, as a separated instance); henceforth, it would be absurd, and even counterproductive, to want to preserve it.

6. However, there are such things as technologies. And these technologies help us escape all sorts of environmental predicaments. These technologies promise us a beautiful future inside an artificial world where they will have demonstrated that there is no limit to our co-*agency*, our partnership with the forms of life that we have yet to decimate and the machines we will have to invent to replace them.

There is no longer any doubt that hypermodernity has turned out to be eco-constructivism and accepted the call for intervening yet even more, continuing its research into a cognitively damaged Prometheus that is quite cognizant that fire burns and yet will call for an even greater increase of industrial fires.

From the ecology of resilience to Latour's "political ecology," eco-constructivism seems to be in a difficult position for pitting itself against geoengineering and its version of the Anthropocene. In his most recent work, Latour seems to grasp the gravity of the danger that climate engineering represents: If we hand over the Earth to "those who Hamilton calls the *Earthmasters* . . . , what a mess they'll make of it!,"[1] as if we were speaking of children who will make a mess and wreak havoc in their bedrooms or a group of individuals who we know will wreck a project and not be capable of properly completing it . . . Actually, for Latour to truly be capable of questioning climate engineering, he should profoundly question his passion for technologies. As Naomi Klein notes, in clearly targeting Latour's thesis: "Love your monsters is a terribly poor metaphor for geoengineering. First, 'the Monster' we are asked to love is not some mutant creature of the laboratory, but the earth itself. We did not create it; it created—and sustains—us."[2] Actually, the eco-constructivists don't believe that the Earth sustains us but that, according to the terms Erle Ellis uses, the Earth is artificial. So, why shouldn't we then, as Gunderson and Holling suggest, become "planetary terraformers"?[3] Modern, hypermodern even, technophiles, adepts of the democracy of the economy (the subordination of politics to the economy), the eco-constructivists produce a theoretical foundation that is completely compatible with geo-constructivism. And yet, there are certain tensions that should be noted between eco-constructivism and geo-constructivism:

1. First of all, we have noted that the ecology of resilience forged its principal concept in direct contrast to the definition of resilience used by the engineers. An ecological thought of resilience can resolutely lead to completely rejecting a project whose aim would be to turn the Earth into a machine dependent on a perpetual technological intravenous drip and incapable of adapting to unpredictable changes—changes that, and this is the leitmotif of the ecology of resilience, inevitably occur.

2. Secondly, a large number of eco-constructivists would clearly reject the extraterritorial position of the geo-constructivists who consider themselves as residing *off-planet*, capable of shaping the Earth as if they weren't themselves residing on it. In complete contrast to this position, the eco-constructivists claim that everything is connected, everything is attached, that there is no place on Earth that isn't disconnected from the rest of it, humans are in permanent contact with nonhumans, who, just like us humans, are also actants. Hence the possible tensions, residing within eco-constructivism, between those who

will tend to merge their approach with geo-constructivism and those who will strive to maintain, through their direct contact with geo-constructivism, a rather healthy difference. We will return at the end of this chapter to a possible extension—and extinction—of these tensions, by paying attention to the theses put forth by Isabelle Stengers and Donna Haraway. We will simply note, for the moment, that the passion through which geo-constructivism produces its magnetic force (or its Frontier magnetism) and that seems capable of attracting pretty much every eco-constructivist—whether they are for attachments or for the great decoupling—is the love of technology, an unconditional love that seems similar to the kind of obligation found within the concept of the superego. It's as if the proclamation that nature doesn't exist was paid for through the following unconscious proclamation: Technology *surexiste*—it *over-exists*—inundating every corner of the planet. This is precisely what eco-constructivists, politically refractory to geopower, must pay careful attention to: The Earth to which the eco-constructivists say they are attached is the one being remodeled by the geo-constructivists.

"In no way am I a technophobe and I believe that we must stop denigrating the instinct to consume."[4] What is worth noting in Latour's pronouncement here is the relation between the two phrases that comprise it: Where the first part of the sentence seems open and hopeful, nonsectarian, the second part of it shows that the absence of any sort of technophobia, within this context, far from maintaining any kind of healthful restraint, is a complete and full acceptance of the industrial mode of the production of technologies and objects validating a model of consumption declared as being "instinctive"—quite ironic for a post-naturalist; a truly instinctive consumption, and in no way constructed by the capitalist economy or by the *designers* of marketing. Actually, for many of the ecomodernists, technology is the site where anthropocentrism has hidden itself, having seemingly escaped through the window of post-environmentalism— the anthropocentrism of the consumption of the world. Used by a privileged number of human beings, technology is the unthought [l'impensé] that makes it okay to write without any hesitation, "Don't let France move toward degrowth." Who made this statement: Bruno Latour? Ted Nordhaus and Michael Shellenberger? Erle Ellis? No. It was a statement made in 2013 by Pierre Gattaz, president of MEDEF (the French neoliberal employers' organization).[5]

FASTER, STRONGER, MORE THAN HUMAN: ACCELERATIONISM, TRANSHUMANISM, AND EXISTENTIAL RISKS

In this light, nothing would prevent us from situating certain tendencies of the so-called "Accelerationist" movement as comprising some of the most powerful elements within eco-constructivism. Whether on the Right (Nick Land) or the Left (Alex Williams and Nick Srnicek), accelerationism rejects—following the ecomoderns—any thought of a withdrawal, restraint, of slowing down, any attempt at protecting or withdrawing oneself from the oncoming socio-economic system. Symbolized by the name Nick Land, the first accelerationist wave made use of the anarchist tendencies of "French Theory" from the 1970s (Deleuze, Guattari, Lyotard), so as to sing the praises of cyberculture and the inhuman forces that it expresses and the deterritorialized virtues of capitalism.[6] Formed in the first decade of the twenty-first century, the second wave of accelerationism is much more in search of a new order, of a reformatted Earth, than the *jouissance* offered up by the extreme universe of machinic delirium.[7] For these new accelerationists, one is better off thinking of *Terminator* (dir. James Cameron, 1984) than the primitivism of *Avatar* (dir. James Cameron, 2009); but the former example should seriously consider entering into politics in order to take care of the environment—a message that the leading actor in *Terminator*, Arnold Schwarzenegger himself, had already perfectly anticipated in becoming a Republican governor of California engaged in helping to fight against climate change . . .

The "Accelerationist Manifesto" by Alex Williams and Nick Srnicek profoundly irrigates this second wave by defining themselves from the very first lines of the manifesto as "modern" and attentive to the "breakdown of the planetary climatic system."[8] For Williams and Srnicek, the only way of alleviating the dangers befalling us is not to oppose the neoliberal economic system that underpins the form of development that has led to innumerable environmental disasters but to accelerate this process (i.e., "liberate the latent productive forces" that capitalism hinders).[9] On one hand, the authors declare: "Never believe that technology will be sufficient to save us"; but, on the other hand, several pages later the same authors call for "a Promethean politics of maximal mastery over society and its environment."[10] For Ray Brassier, anyway you put it, this mastery is ontologically inevitable, the critique of Prometheanism only rests on reality by way of a refusal staked on admitting that there is no "natural" given or cosmic "equilibrium": Metaphysically correct, "Prometheanism is an attempt to participate in the creation of the world without having to defer to a divine blueprint."[11] Far from representing a heretical tendency, the position held by Brassier is clearly in line with the paradigm of turbulence in justifying the inevitability of the reprogramming of the world under the pretext that everything is a "dynamic process" and is "perpetually being redetermined."[12] That Marx is also evoked by Brassier in order to emphasize "re-engineering ourselves and our world on a more rational basis"[13] indicates quite well the crux of the problem: Where Marx would consider that the instance of transformation was politics, for the accelerationists and their fellow travelers, politics must cede its place to technology and technology's power of recombination. Established on the belief in a lifesaving technology, the "Promethean" politics of the accelerationists opts for an Anthropocene of the great acceleration.

So we will not be surprised to see Williams and Srnicek dreaming of "the quest of Homo Sapiens beyond the limitations of the earth and our immediate bodily forms," wishing for finally achieving delivery on "the promissory note of the mid-Twentieth Century's space programmes."[14] Indeed, nothing can truly prevent a movement of theoretical acceleration sufficiently powerful enough for breaking free from terrestrial gravity as well as the weight that holds down bodies and ties them to their "immediate bodily form"—that is to say, their biological finitude. And yet, it's indeed the hatred of this finitude as with every limit that links the accelerationists with the transhumanists, those theoreticians who envision how evolution will soon lead us to a position where humanity is left behind as an outmoded species. According to the transhumanists, cybernetics, nanotechnologies, and genetic engineering will allow for the production of a more improved human being, a transitional human (which is what *transhuman* literally means for its promoters) that will end up giving way to a truly posthuman (i.e., a life-form that—by way of its cognitive capacities and its longevity—will supplant humans just as humans supplanted primates).[15] As André Gorz notes in his lecture on a "posthuman civilization" promised through an alliance between a science cut off from the sensible world and a capitalism that cuts individuals off from their agency and knowledge, what these projects that privilege the future of artificial intelligence reveal is, in fact, a "hatred of nature and life." And this is how Marvin Minsky, a world-renowned specialist of artificial intelligence and cognitive sciences, can make his claims of "a contempt and disgust for this meat machine that is the brain and the bloody mess that is the human body."[16]

In the same way the accelerationists integrate into their position questions about climate change, the perspicacious transhumanists take into consideration the fact that the evolutionary possibility that could lead us—via the transhuman—to the posthuman, even to the singularity (this artificial intelligence that could supposedly emerge thanks to the considerable developments in computational capacities of computers connected to the Internet) could be seriously compromised in case of the premature disappearance of the human species[17] . . . This is why the philosopher Nick Bostrom—one of the thinkers originally involved in the transhumanist declaration that has already been rewritten a number of times,[18] cofounder of the World Transhumanist Association, and director of the Future of Humanity Institute (at the University of Oxford)—is insistent on reminding us of the "existential risks" that currently threaten humanity and its entire ecosphere. An existential risk, he writes, is:

> the premature extinction of Earth-originating intelligent life or the permanent or drastic destruction of its potential for desirable future development.[19]

Nick Bostrom points out that in our current times, the principal existential risks are no longer natural but anthropogenic, resulting from our capacity to "radically expand our ability to manipulate the external world or own biology" via biotechnology, molecular nanotechnologies, and intelligent machines.[20] In this light, it is interesting to note that Bostrom is an adviser for the Center for the Study of Existential Risk at the University of Cambridge , a center founded by the philosopher Huw Price, Jaan Tallinn (cofounder of Skype), and Martin Rees (a scientist whom we have already mentioned earlier).[21] The focus of the center is to forge a science that would be capable of reflecting on the dangers of specific technologies that could exceed our control and eliminate us. That a scenario that up until now would have simply been found within the industrial imagination of Hollywood has led to the creation of such a research center counting among its members some of today's most important scientists (such as Stephen Hawking and Peter Singer, among others) is, at the very least, a bit disconcerting—all the more so when we know that one of its founding members, Martin Rees, has opted for the Plan B of climate engineering. This *double bind* can therefore be analyzed in the following way: All these scientists and philosophers have perfectly understood the bio-existential risks that technologies pose to humanity (in passing, the animal question doesn't seem to theoretically or affectively embarrass them); but—such is the imperative of the von Neumann Syndrome—they are unable to justify the possibility of suspending a technological innovation. Withdraw? Never!

Nevertheless, a small doubt begins to claw its way into the conversation: Transhumanism seems rather different than ecomodernism, and there is no guarantee that an ecology, albeit an ecology that is post-environmentalist, would be on the side of the accelerationists. And yet, the post-environmentalists Shellenberger and Nordhaus don't hide their lively interest in the transhumanist trend and have done an interview with one of these sycophants, the sociologist Steve Fuller from the University of Warwick.[22] During the course of their interview, Fuller explains that we should abandon any idea of the precautionary principle, which is based only on fear, and take up the "proactive principle," which turns risks into "chance." Fuller defends the necessity for life-size scale experimentation and declares that the priority must be focused more on the increase of the power of Humankind than on the protection of nonhumans. As an indication of this, we should note that Fuller is also known for his defense of the theory of intelligent design, a theory that resuscitates creationism under the guise of scientism.[23] The integral design of the Earth is without a doubt the point of convergence for transhumanism, accelerationism, and of an eco-constructivism, swallowed whole and digested by geo-constructivism.

COSMOTECHNOLOGIES AND CONSCRIPTED APPARATUS (SEPARATING TECHNOLOGIES)

If it has been shown that the technological question is at the heart of the way in which the accelerationists, the ecomodernists, and the post-environmentalists relate to thought and the practice of ecology, then it is high time to take up this question at arm's length and posit the foundations of a politics and a *non-constructivist* relation to technologies. What is, in fact, the trap set by the eco-constructivists? Reducing thinking concerning ecology to having to choose between two possibilities: either a hatred of technology or its unconditional love. Either reject all technologies without any exceptions or accept all technologies. Either refuse all development or consider those who refuse development as "stupid." Our objective is not about finding some sort of middle-ground position but more specifically about locating a political position whose aim would be to *provide room for distinguishing* between what we want and what we don't want, in order to radically contest not only eco-constructivism, from the ecology of resilience to the accelerationist approaches, but also the myriad of geo-constructivist discourses analyzed earlier. An *ecology of separation* must, first of all, teach us how to distinguish between technologies—but what are the criteria that could help us to evaluate and separate the various technologies? In his book *La domestication de l'être*, Peter Sloterdijk distinguishes between:

1. On one hand, "allotechnics," which lend themselves to the practice of raping and pillaging and end up leading to the "destruction of primary materials." These technologies are always applied from the outside; they are exercised by a subject (a master) who applies his power over an object (a subordinate) who he seeks to control.[24] As certain thinkers of resilience have understood, it's the very fact itself of wanting to control a supposedly predictable nature from the outside that renders ecosystems vulnerable and leads to an "erosion of resilience."[25]

2. On the other hand, "homeotechnics": Elaborated through the paradigm of information, through the "theory of complexity," and ecology, all these technologies imply a strategy of "cooperation," of a "dialogue" with nature. Sloterdijk writes that homeotechnics "can't want anything that is entirely different from what the 'things themselves' are or what they can become by

themselves."[26] We should note that André Gorz takes back up these same themes at the end of one of his late works, *The Immaterial*.[27] Borrowing from the works of Ivan Illich, Gorz distinguishes between "locking technologies" that lead to the domination of nature, that "dispossess men from their living environment," and "open technologies" that "favor communication, cooperation, interaction, such as the telephone or today's networks and open source software."[28] The examples Gorz refers to should be contested: a cellphone is perhaps the best way to lock onto someone within their trajectory (i.e., to control and practice surveillance over individuals); in a similar manner, networks can constitute a new space of enclosure. But, nevertheless, let's try to grasp the following point: What both Gorz and Sloterdijk clearly highlight is the opposition between cooperative technologies and self-enclosed technologies, the latter leading to what Illich called a "radical monopoly," which "substitute themselves for what individuals need to do for themselves."[29] Today this radical monopoly is nothing other than that which makes a democracy of the economy possible—in other words, the possibility of an unbridled capitalism.

However, we can immediately see the problem posed by the ecology of separation we are beginning to sketch: How do we denote the difference between good and bad technologies without presupposing a predetermination of technologies that won't take into account the manner in which they are used? The manner in which they are used can, in turn, have the capacity of modifying the technologies in the midst of their individuation—that is, in the midst of their improvement.[30] Are not the uses of technologies variable and unpredictable? Will the critique of the end of the Great Divides and of a generalized interconnection not lead us toward a new essentialism? In fact, it is actually necessary to think together both the technologies and the world they tend to produce (or destroy): every technology is a cosmotechnology. If, for example, nuclear power is an allotechnics, a locked technology, it's because nuclear energy requires a secret (that of having to transport and deal with nuclear waste), an army to defend it, as well as an economic and policing apparatus that allows for this self-enclosed technology to exist. It's not because there is an essence of nuclear power but that there are *apparatus called forth and conscripted* by this technology. A conscripted apparatus is doubtless not inscribed within the essence of a technology, but it is the complement without which a use of the technology would be impossible or suicidal (such as publicly indicating to potential terrorists or antinuclear activists the location of a train carrying nuclear waste . . .). In this sense, a technology and its conscripted apparatus produce a certain type of *world* (of society, of an individual relation with energy, to energy consumption, etc.). To believe that nuclear energy could vary in terms of the way it is used, to believe, for example, that nuclear energy is possible without any army is to believe in a very dangerous and idealistic fiction—the constructivist fiction of worlds without limitations that we can construct at will.

In the end, it is only by way of envisioning problems from a cosmotechnological viewpoint that we will be able to 1) discern a position between values and desired worlds; 2) know, in terms of causes, what the called forth and conscripted apparatus by this or that technology are and the type of world associated with them; 3) to decide a politics that privileges technologies that are *open source*—and here again, it's by way of the action of separation (to provide a position) that presides over relation (openness, cooperation).

AGAINST THE INCONSEQUENTIAL POLITICS OF ECO-CONSTRUCTIVISM

And yet, eco-constructivist politics has something entirely different in mind than privileging open technologies and the forms of life we could associate with them. In an interview that he gave to the Breakthrough Institute, Steve Fuller uses as a reference the analyses of the transhumanist named FM-2030 in order to proclaim a great turn in the way in which we envision political polarities: It's no longer the opposition of the left/right that have brought about the structuring of political space but the opposition of high/low. Toward the high end are those who believe in permanent technological innovation and the absence of natural limits; toward the low end are those who irrationally support the opposite thesis. However, according to our sociologist, the left is divided between, on the one hand, "communitarians" (here we can think of the "old" environmentalism that has been thwarted by the post-environmentalists, and Fuller more than likely includes in this category the Zadists,[31] the anti-industrial movement, those opposed to fracking in defense of certain natural lands and waterways, etc.), and, on the other hand, progressives (those Fuller calls "technocrats" who claim to be ecomodernists and, undoubtedly, pro-accelerationists). In the same way, and this is still according to Fuller, the right also has its "traditionalists," on the one hand, and, on the other, its "libertarians." Fuller calls for the technocrats and libertarians to join forces against the traditionalists and communitarians.

It's precisely these latter that Williams and Srnicek lash out against in their accelerationist manifesto. They claim that the current left is divided between those who cling "to a folk politics of localism, direct action, and relentless horizontalism" and who are "content with establishing small and temporary spaces of non-capitalist social relations," valuing "authenticity," the "common," "neo-primitivist localism," and "communal immediacy."[32] Are Williams and

Srnicek chastising the left for not being revolutionary? No, since, as we have seen, they maintain that we shouldn't get rid of neoliberalism and capitalism but rather push them to their limits and fight them on their own terrain. But, on its own terrain, capitalism always comes out the victor—with the destruction of the very terrain. In the end, this has little importance for either Williams and Srnicek, the ecomodernists, or the post-environmentalists, since what truly matters is criticizing the left and, more specifically, the ecological or environmental left, whether they are supporting the stance of de-growth or as inheritors of deep ecology (Latour's targets), against nuclear energy or fracking, and, in a more general manner, antiextractivists (we would do well to recall the attacks made on them by Ellis, Nordhaus, Shellenberger, Lynas, and Brand) or whether the left is anchored in the defense of territories (a defense that is completely intolerable for Williams, Srnicek, and Fuller). The politics of eco-constructivist politics is what motivates the lords of the times: neoliberal, capitalist, technophile, global, and unabashedly hypermodern. Under the pretext of a turbulent world, full of "surprises," the eco-constructivists leave causes in the hands of industry and take as their responsibility the administration of the consequences. It's this administration a posteriori that they call "ecology," "modernity" being the act of the prince—and the principle—of production. Within such a schema, resilience is the vain hope of nondisastrous consequences. Alas, this hope does not follow the rising rate of CO_2 emissions or the increasing rate of climate engineering projects—hence its *inconsequentiality*.

The politics resulting from an ecology of separation implies knowing where to evaluate the opposition: not from outside the hegemonic space of unrestrained technologies but from within the interior of this space. The resilience that we are proposing has as one of its conditions the possibility of an internal or interior distance, an outside within the inside of the world itself. It's precisely this outside that doesn't exist for the constructivists under the pretext that everything is interconnected. And yet, it is indeed possible, as we will attempt to show in the third part of our book, that interconnection requires this internal separation, an immanent distance, an outside without which the notion of an inside has no meaning whatsoever. In other words, we can perfectly understand the critique made by Williams and Srnicek relative to the fetishism of a direct democracy and of a horizontality in regard to certain current political struggles; but this critique is what allows the two authors to position themselves within an outside that is absolute, transcendent, a *distancing without return* that places them, just like their geo-constructivist cousins, outside of the world. We must, therefore, avoid just as much the extraplanetary position as that of a pure immersion into the world; in other words, we must avoid the geo-constructivist excess that is also shared by ecomodernists (Ellis, Williams and Srnicek, Brand), as well as the eco-constructivist excess, who only swear by networks, relations, and attachments. An ecology of separation produces detachments from the interior of the world and distant viewpoints that it repatriates within territories. It is thus perfectly suited for supporting all the territorial and antiextractivist struggles when these struggles open their identities up to all the differences that populate the world—the common world that the minoritarian bodies of the Anthropocene inhabit and defend.

THE FUTURE OF THE CHTHULUCENE: SLOWING DOWN, MAKING KIN, COMPOSTING — BUT WHAT ABOUT NAMING THE IMPOSSIBLE?

To take back up the title from the second part of this book, what is the "future of eco-constructivism"? To respond to this question, it is somewhat necessary to develop what we evoked earlier on: the tensions inside eco-constructivism between those who have been able to create a rather powerful resistance to the magnetic force of geo-constructivism and those who have been unable to do so.

1) It's quite clear that a thought such as that of Isabelle Stengers is well equipped against the geo-constructivist economy. Defending the necessity of a political ecology, Stengers insists on the incompatibility of this political ecology with "capitalist logic," a logic that is "intrinsically incapable of being civilizable because what matters for it is not possibilities for relations, but opportunities for exploitation."[33] Hence Stengers's appeal for a *slow science*: "Speed demands and creates an insensitivity to everything that might slow things down: the frictions, the rubbing, the hesitations that make us feel that we are not alone in the world" to the possibilities of "thinking and imagining, and in the process, creating relationships with others that are not those of capture."[34] So here you find us, resolutely positioned a long way from any sort of accelerationism;

2) In a similar vein, the recent essays by Donna Haraway consider the Anthropocene as a limit-event—rather than a new era—that we must integrate into the longer age of the "Chthulucene," the era that has long been underway where multiple temporalities, both human and more-than-human forces, as well as forces other than human or nonhuman, have become entangled, names of the Earth of which Gaia is only one example among many (Naga, Tangaroa, Terra, Haniyasu-hime, Pachamama, A'akuluujjusi, etc.).[35] Inside the Chthulucene, the pioneers of terraforming are mere bacteria. This sort of transhistorical perspective—i.e., a perspective that includes the most distant

prehistory and that also projects us into the furthest confines of human possibilities—resituates the Anthropocene not as the summit of anthropomorphic creativity but as a horrendous ecological, social, and cultural impoverishment: a planetary ecocide seriously endangering the capacity of both human and nonhuman systems for regenerating themselves. Relying on the work of Anna Tsing, Haraway identifies one of the nodal losses that could be marked by the Anthropocene limit-event, understood here as an unprecedented catastrophe, a rupture and not simply a duration: the eradication of sites of refuge where human or nonhuman populations are capable of reconstituting themselves after materially traumatizing events (desertification, clear-cutting, etc.).[36] From this point forward, the question becomes: How can we figure out how to make it so that the Anthropocene does not endure—that is, so as to prevent it from installing itself as an era? The response that Haraway gives, and that will get all of our attention, is the following: "Make Kin Not Babies!"[37] Rejecting the facile critique of the left that would see in such a slogan the hardly hidden hand of a neoimperialism, of misogyny and racism, Haraway insists on the material urgency of the slogan: An overpopulation of humans will be disastrous for both humans and nonhumans.[38] Far from privileging technologies (motors, tools, machines), Haraway's analysis invites us to bring into existence new technics of kinship, a new terrestrial cosmopolitics aiming to re-establish the way in which we symbolize our kinship links, whatever the nation or species in question: "I'm a compost-ist, not a posthuman-ist: we are all compost, not posthuman."[39] Where posthumanists can't master their passion for technologies, these technologies are simply used by compostists for accompanying the conversion of organic materials.

Haraway's Compost-ist Manifesto and the antiaccelerationist temporality proposed by Stengers both, in a certain manner, resonate with postcarbon, de-growth perspectives or certainly resonate with the transition town movement, with which our ecology of separation has affinities. However, it seems to us that such perspectives require not only to construct and compose new possibles but also to recognize a part or place for the impossible—*the insistence of the impossible*.[40] Indeed, it's the impossible that insists to the point of demanding of us, urgently, to decelerate and to create a Composters International. Indeed, it's the presence of an "inhuman nature" (Nigel Clark) that has arrived to shatter our fantasies of symmetrical compositions where human subjects and nonhuman objects would be endowed with the same quantity of ontological and political power.[41] At once terrorizing and splendid, the monster that is the Chthulucene takes a sinister pleasure in returning eco-constructivism back to a local assemblage, always under the threat of a tidal wave or a volcano—these manifestations of nature that are beyond any possible "negotiation."[42]

Indeed, it's through experiences and events where the nonhuman loses its superficial skin as object or actant, no longer being what we *"make act"* (Latour) and revealing its somber and inhuman center where action and its social dramas transform into tragedies. This worrisome depth imposes a radical asymmetry between the power that it is able to exert and the excess that such a power evokes for humans. This excess cannot be contained very long within the eco-constructivist framework. Having arrived at the end of its theoretical journey, eco-constructivism must dissolve into the furious ocean of the unconstructable.

PART III

AN ECOLOGY OF SEPARATION

Natured, Naturing, Denaturing

If geology, or the "mining process," opens onto an ungroundedness at the core of any object, this is precisely because there is no "primal layer of the world," no "ultimate substrate" or substance on which everything ultimately rests.

—Iain Hamilton Grant, "Mining Conditions: A Response to Harman"

OBJECT, SUBJECT, TRAJECT

Throughout the first two parts of this book, we have examined the price to be paid for getting rid of any idea of something we call nature. This price is *exorbitant*, in the proper sense of the term: We have to get ourselves to exit any orbit, propelling ourselves into a stratosphere, one that is indeed much more imaginary than real, of an unbridled construction where humans are agents of mastery, into a stratosphere of limitless developments and technological monsters that should deserve our unconditional love. The absence of nature legitimizes the fantastic possibility of remaking the world in order to steer it, to be its pilot, to manage it; but the world, inevitably, withdraws from the human setting, leaving the latter alone—without nature and without a world.

How can we avoid both this anaturalism and this absence of a world, this *acosmism*? How can we propose an approach to nature that isn't "ideological," nostalgic, or reactionary—namely, how do we resist the *naturalist* temptation? This naturalist temptation will be the name we propose for the translation of the desire to posit a unilateral affirmation of the power of nature—against the geo-constructivists as much as the eco-constructivists—a nature that would be an omnipresent substance that would contain everything in its core and thereby condemn us to an absence of any outside—that is, to the cancellation of any existential or political alternative. If it is necessary to oppose the reduction of a nature to a devitalized object, it appears that positing nature as an *absolute subject*, as naturalism tries to do, inevitably leads to a form of symmetrical acosmism with devastating constructivist effects. How can we find a way out of this impasse? By affirming our relational condition without succumbing to a pathological connectionism, to the contemporary passion of indivisible links that—despite their appearances—unite two sworn enemies such as deep ecology and eco-constructivism. By simultaneously recognizing our being-in-relations and the demand for an outside.

In order to defeat this *transcendental narcissism* of the human being—namely, the fiction of a humanity hallucinating itself everywhere, on Earth as much as in the sky—it's not enough to simply affirm nature as subject, as a number of thinkers and ecological and environmental activists do, concerned with shielding themselves from the reification of nature; it is also necessary to show that nature—the nature we will refer to as *real nature*—is something that exists as separate from thought—separate from thought, but not split from it. Even though nature is revealed to us through thought, nature remains secluded, closed in on itself. We've posed a great number of questions to it and have made it speak much more through force than through reason, and nature perseveres in not answering us. Our inquiry will attempt to identify this mutism, this gray area of nature that we will call its *unconstructable share* [part inconstructible]. This is the share that neither Spinoza nor Arne Næss seem capable of seeing, an aspect that Whitehead seems to discern but that is nonetheless only clearly theorized by Schelling: Nature is not simply natured nature (an object to be shaped or that is manipulable), naturing nature (a producing subject), but also a *denaturing nature*—a movement of withdrawal, an antiproduction preceding all production.

It's through this foundation of such a conception of nature that we will reexamine our relation with the Earth. Neither object (as the geo-constructivists would like it to be) nor quasi-subject (as the neo-organicists would like it to be), the Earth is first and foremost a *traject*, a long-term trajectory originating from out of the depths of time and destined for extinction.[1] Far from being compact and malleable, the Earth is opaque. Much to the chagrin of the human being, it does not provide a full and clear image of itself: will humans be capable of inhabiting an Earth that withdraws from a project of integral constructability?

CHAPTER 10

NATURING NATURE AND NATURED NATURE

NATURE CUT IN TWO

Following a philosophical tradition that seems to go at least as far back as the thirteenth century, we will maintain that it is ontologically useful to distinguish between, on the one hand, naturing nature (natura naturans)—that is, the permanent genesis of things, nature as a process, as productivity—and, on the other hand, natured nature (natura naturata)—that is, created nature, finite objects that are the result of this process.[1] The question we would like to respond to is the following: For the ecology that we would like to establish, is it preferable to maintain, and even effectuate, this difference or to remove it?

In fact, this distinction was not born in the thirteenth century; we can already discern it in the work of Aristotle, for whom nature is "the source from which the primary movement in each natural object is present in it in virtue of its own essence."[2] For Aristotle, the movement in the sublunary world is not a state but a process: a being becomes complete, becomes actualized, becomes—through changing—who it is. This changing-in-movement has an *end*—that is, simultaneously a goal (*telos*) and a completion. Every becoming veers toward a final immutability, which is not the immobility of that which would be unable to be moved—one shouldn't confuse the final unchangeable position of rest with an initial immobility that would be foreclosed from any becoming (from any "life" if you will).[3] All the other meanings of the word *nature* are, according to Aristotle, subordinate to this principle: Matter can also be considered as natural, but it's only because it can "receive" this principle and develop its potency; it is the same case with form (eidos)—that is to say, the being considered once it has become achieved, complete. Before being some sort of object in the midst of formation, or formed, nature is the principle of growth and immanent transformation. Immanent in the sense where the natural process sets about from itself—and not from an external action of a human being, of his art (techné)—and spreads out in a direction from itself, in the sense where the process of growth does not have its sights set on something that would be artificial.[4]

During the Middle Ages, this Aristotelian difference between nature understood as a principle of growth and as immanent transformation and nature as a finite object (that is formed, completed, achieved) was—as Frédéric Manzini states—"over-interpreted in a religious sense."[5] This overinterpretation inadvertently consisted of overdistinguishing, on one hand, the *natura naturans*, the principle of growth becoming the manifestation of the creative capacity of God, and, on the other hand, the *natura naturata*, what is created. The irony of this story is that the distinction *natura naturans/natura naturata* is still, up to this very day, known through the reprisal of the terms by Spinoza, whereas even his reprisal of the terms consisted of doing away with the theological distinction of the terms. In fact, for Spinoza, two ontological operations are required: a) referring to naturing nature as God and *connecting* both of them; b) fusing together the naturing and the natured. In his *Short Treatise*, Spinoza recognizes the Thomists for having "understood God through" the *natura naturans*; nevertheless, he adds, "their natura naturans was a being (as they said) beyond all substances."[6] Within the counterproductive interpretation that Spinoza proposes of the scholastic distinction, the productions of nature participate *without a solution of continuity* in the productive power that never abandons them, this power remaining substantially present at each stage, each level, and each site of a permanent production.

Why did Spinoza judge it necessary to identify God with nature—*"Deus, sive Natura"* as is stated in his *Ethics*[7]—and affirm the continuity of the naturing and the natured against its disjunction? A priori, for some very good reasons, for reasons that could make us want to proclaim him as the prince of a metaphysics of ecology: Because he saw the calamities of this disjunction once this disjunction is resumed, translated, and reinforced—and not refuted—by the science of the seventeenth century. Indeed, if all the power of creation resides with God, then everything that is excluded from this power will literally be *what remains*—amorphous, ready to be mathematized and controlled. For Descartes, movement and rest will be disconnected from change understood as creative metamorphosis: Movement will have to be studied in and of itself, as a movement-of-an-object and not as a being-changing-in-movement (as with Aristotle). Suddenly, nature will become pure natured, *finite* in a new sense: not conscripted to blossom and come to a completion in a form (as is the case with Aristotle) but *finished*, done for—lacking any internal force, lacking the capacity to respond to itself, by way of itself; and, in this sense, dead. Must we from then on accept the invitation offered by Spinoza and fuse together the nature naturing with the nature natured? Is this the ontology required for a contemporary ecology? The only way of protecting nature against *physicides*?

THE PROGRAMMED FAILURE OF THE SPINOZIST CURE

Spinoza's philosophy could appear as the best possible remedy against Descartes's anthropocentric and technophile anaturalism, as well as its geo-constructivist extension. Where Descartes would be the symbol of the obstruction to an ecological thought and its cause, or at the very least, one of the *discourses on method* leading to environmental disasters, Spinoza would be the promise of an ecological ontology capable of overcoming the fatal separation between humans and nonhuman nature established by Descartes—to the first, reason, thought, speech, and free will; to the nonhumans, automatism.[8] It is no doubt for this reason that Arne Næss made Spinoza's philosophy one of the foundations of deep ecology, deep ecology's task being in part to heal a supposedly sick ecology of its "superficiality" (its inability to critique anthropocentrism). Of course, as David Rothenberg points out during a conversation with Næss, Spinoza wasn't talking about "saving the earth," but Spinoza's philosophy forces us to consider "our direct connection with all things around us. It's an early plea for an ecological vision, . . . the inspiration for ecosophy."[9]

Næss confirms this throughout the entire conversation with Rothenberg: Yes, Spinoza can help us to undermine the "cleavage between the spirit and the body, and therefore also between subject and object."[10] For proof of such an undertaking, we will cite the famous passage from the *Ethics* where Spinoza affirms that the human being, far from being definable by some sort of free will, is not "a dominion within a dominion" and "follows" rather than "disturbs" the order of nature.[11] Is this not a philosophy able to contest the dominant discourse of the Anthropocene? Infinitely fertile, nature would not be separated from God, and this lack of separation would lead to the inability of placing a creator God on one side and, on the other, a nature that would be, from then on, sterilized.[12] Thanks to Spinoza, nature would have rediscovered its dynamism: The effort (conatus) of each thing to persevere in its being, in order to oppose what would tend toward effacing its existence, and so as to increase its power of being, would express the vital grandiose affirmation that traverses creation.[13] Have we not uncovered here a marvelous naturalist medicine!

However, let's take a closer look at the so-called ecological promise of Spinozism. First of all, we should be a bit skeptical about the veritable capacity of this philosophy for escaping an anthropocentric framework since the *Ethics* also affirms that while animals "have sensations," this in no way prevents us, in the name of the search for what we would deem as "useful," to "use them at our pleasure, and treat them as is most convenient for us": It is clear, Spinoza writes, animals belong to those entities "whose nature is different than human nature."[14] How can we explain this *return of dominion*, this anthropocentric asymmetry—in other words, this difficulty in taking into account nonhuman animals? To answer this question, we must plunge into the heart of Spinozist ontology and the interpretation of Spinoza that is dominant today, which characterizes Spinoza as the thinker of experimental passions and profound individuations. Deleuze largely contributed to this fictitious Spinoza considered by way of the "middle," by the *"common plane of immanence* on which all bodies, all minds, and all individuals are situated"; but in order to directly begin with the modes, the relations of speeds and slowness and affects, one must—as Deleuze clearly states—get rid of the "unique substance,"[15] dry out the "the continuous line or tide of propositions, proofs and corollaries" that "amounts to a sort of *terrorism of the head* [our emphasis]," and only conserve the other part of the *Ethics*, that is "discontinuous, constituted by the broken line or volcanic chain of the scholia."[16] Once again we have the *pretend as if*, proceeding by way of a negative hallucination, paid for by a horrendous return of substance: The head takes its vengeance, irrigating the plane of immanence on automatic pilot, in its management mode and imposing over all the bodies its infinite appetite of connection.

Ontological immanence is not without its virtues. As Epicurus and Lucretius before him, Spinoza courageously confronted and was against alienating forms of religious transcendence, against dogmatic subjugation that tends to replace the literal reading of books by a simple blind faith.[17] But this grandeur can also contain rather strict limits that run the risk of inverting the meaning. For, in Spinoza's philosophy, immanence is absolute, the verb *im-manere* is taken literally: to remain (manere) in oneself (in) without anything ever escaping. "What defines an immanent cause is that its effect is in it—in it, of course, as in something else, but still being and remaining in it."[18] It's in this way that, as Merleau-Ponty also writes in regard to Spinoza, the infinite "contains" the finite—that encompasses it, limits it, and forces it to merely "[draw] on the finite from out of the potency of an infinite being"[19] Contained in such a way, creation never escapes from its creator: "Whatever is, is in God, and nothing can be or be conceived without God," writes Spinoza.[20] The system of absolute immanence runs the risk of leading to what Hegel called an "acosmism":

> Nature, the human mind, the individual, are God revealed under particular forms. It has been directly remarked (pp. 257, 258, 280) that undoubtedly

Substance with Spinoza does not perfectly fulfill the conception of God, since it is as Spirit that He is to be conceived. But if Spinoza is called an atheist for the sole reason that he does not distinguish God from the world, it is a misuse of the term. Spinozism might really just as well or even better have been termed Acosmism, since according to its teaching, it is not the world, finite existence, the universe, that reality and permanency are to be ascribed, but rather to God alone as the substantial. . . . According to Spinoza what is, is God, and God alone. Therefore the allegations of those who accuse Spinoza of atheism are the direct opposite of the truth; with him, there is too much God.[21]

But we can understand the term *acosmism* in two different ways:

1. The first way consists of accusing Spinozist substance of only allowing for what isn't it to remain by way of annihilating it: The passage that leads from infinite substance to finite things would be of the order of a degradation, of a decline. Defended by Hegel, this interpretation is also supported by White-head: Spinoza "bases his philosophy upon the monistic substance, of which the actual occasions" (i.e., the name Whitehead uses for defining *becoming* in its singular, punctual becomings) "are inferior modes."[22]

2. However, Pierre Macherey attempts to show that this interpretation was untenable: Within Spinoza's philosophy, the infinite and the finite do not de-scribe two separate orders, it's completely the opposite, there is "but one single and same continuous and indivisible reality."[23] This is true—but the entire problem rests precisely on this continuity, this indivisibility, this absence of separation and of loss. On the plane of essences, infinite substance is affirmed everywhere in an equal manner; but when it becomes a question of thinking existences, what is imposed is a brutal mechanism: Each thing proves to be determined from the exterior by way of the others with which it is assembled, by way of the mechanical action of what Martial Gueroult calls (in his analyses dedicated to Spinoza) the "pressure of the ambient forces."[24] Too much God, on the one hand, and too much mechanism, on the other—and no world be-tween the two of them. Of course, Spinoza is right, each thing experiences the pressure of what surrounds it; but this pressure doesn't allow for the singular individuation of beings, their existence, which cannot be reduced to a sub-stantial cause or to a mechanical effect. For each being, to exist means to lose the substantial Infinity, to accept the lack of a guarantee of the whole, without actually losing the possibility of being autonomous and of recasting infinity in the forms of the good, the beautiful, and the true.

Wait a minute. How can this be? We must have gone a bit too far here . . . Spinoza is supposed to be *the* thinker of Nature! For Deleuze, Spinoza proposes a "new naturalism"[25] counter to Descartes. And yet, what is the nature that is at stake within this new naturalism? According to Deleuze:

> It should be clear that the plane of immanence, the plane of Nature that distributes affects, does not make any distinction at all between things that might be called natural and things that might be called artificial. Artifice is fully part of Nature, since each thing, on the immanent plane of Nature, is defined by the arrangements of motions and affects into which it enters, whether these arrangements are artificial or natural.[26]

For Deleuze, nature is nothing more than the plane of immanence that allows us to arrive all the way to the end of the "Spinozist critique of Cartesian dualism" where body and mind [esprit], nature and artifice, become united and indiscernible. Let's say: mixed together. To the point where the concept of nature becomes useless. To the point where we end up being suspicious of detecting a paradoxical anaturalism at the heart of Spinozism: Acosmism produces an anaturalism even in spite of Spinoza's pretensions of speaking about a naturing nature and a natured nature. "What substance is missing," Simondon writes in regard to Spinoza, "despite the Spinozist terminology, is being nature."[27] It's perhaps the definitive welding together [soudure] of naturing nature/natured nature that has liquidated any possibility of existing for the beings that populate the world, an asphyxiating welding just as dangerous as the split it sought to fight against. The promise of a Spinozist ecology turns into a nightmare.

FROM *DEEP ECOLOGY* TO *SHARP ECOLOGY*

Has *deep ecology* escaped from this nightmare? What we know for sure is that Næss is fully aware of the danger that potentially resides within his ecosophy: The necessary fight against "the abyss of atomic individualism" should not lead us to fall into a symmetry—namely, "mystical and organic views" where the expression "drops in the stream of life" runs the risk of succumbing to the "stream of life," where "the individuality of the drops is lost in the stream," this life that Næss, nevertheless, tell us is "fundamentally one."[28] And yet, from our vantage point, Næss's unilaterally relational thought doesn't seem to avoid this danger. Indeed, it's crystal clear that the ontological fight of contemporary ecological thought must be waged on two fronts:

1. On the plane of the status of the subject-object, it is obviously necessary to refrain from abandoning nature to the status of an object that a human subject could control: Nature is not simply a product, it is an autonomous force of production. This is where we must critique geo-constructivism and its Cartesian backing [support], the anaturalist position that definitively reduces nature

to its capacity of resisting the constructions to which we subject it. If nature must be thought of as much more than a simple *finite object*, the human being must strip himself of his imaginary status as infinite subject—the status of an immortal that believes he can escape environmental disasters. And so here finally we encounter what an ontology of ecology must emphasize: the realization and awareness of nature as a genetic power.

2. But this apparent critique of a generalized Cartesianism is not enough. If this critique has no problem confronting geo-constructivism head-on, it nevertheless leaves intact a part of eco-constructivism that suffers from an inverse excess of geo-constructivism: not from the division of the human subject/nonhuman object but from the inevitable interrelation that leads to the inability to gain a perspective, through achieving some distance from everything. What is at stake here is not the *status* of humans, nonhumans, and nature, etc., but rather their mode of *distribution*—of relation or of nonrelation (division). And yet, just like the geo-constructivists, *deep ecology* maintains that there is an absolute interrelation between everything. Here we can see the profound irony that, within the shadows, governs the eco-constructivist relation with the *deep ecologists*: their very large (and very repressed) resemblance as far as the modes of the redistribution of beings. It's true, there is no doubt that in contrast to any sort of geo-constructivism, deep ecology will attempt to grant precedence to the nonhuman world over the human world; but it does so in the name of a Nature that is hyperconnected, "a network"—writes Næss—or "field of relations in which things participate and from which they cannot be isolated."[29]

It is this very difficulty of separating or isolating society from nature that explains the revulsion experienced by Murray Bookchin vis-à-vis biocentric philosophies, accused of covering over precise analyses of the social causes of environmental disasters within the concept of the human species. Brimming with paradox, these "antihumanist" philosophies accuse the human species of all the evils of the Earth to such a degree that they end up declaring the species "antinatural."[30] We certainly follow Bookchin when he describes, well in advance of Malm and Hornborg, the way in which "a sort of human species has replaced the classes";[31] but we are, from now on, confronted with a theoretical problem that Bookchin's invectives against deep ecology don't really help us with:

—Either the problem of deep ecology is that it maintains the great nature-culture divide. This would be its ontological error: It believes itself to be fighting against the destructive forces of the anthropogenic machine; but it has always and continues to situate the human species outside in relation to nature and considers the human species as "antinatural" and, as a result, simply reinforces the roots of the problem itself—i.e., the Cartesian extraterritoriality of placing the Human outside nature;

—Or we must consider that deep ecology refuses this divide: Deep ecology strives for the reintegration of humans into nature. When the *deep ecologists* that infuriate Bookchin lash out at humans, it's not in order to reinforce the great divide but in order to reject the surplus of humanity so as to affirm a natural totality. As we know, this leads to one of the most controversial points of Næss and Sessions's political platform:

> The flourishing of human life and cultures is compatible with a substantial decrease of the human population. The flourishing of non-human life requires such a decrease.[32]

There is indeed a variety of deplorable ways for ridding ourselves of this surplus of humanity: a patriarchal control of the state in regard to the birthrate, for example;[33] or worse, the virtues of an epidemic, of some sort of natural disaster capable of drastically resolving the problem of overpopulation . . . However, these political missteps and this ignorance of human alterity—in spite of Næss's lucidity[34]—appears to us as an effect of a profound ontological impasse within deep ecology: It seeks out the unity, the immanence of the *bios* (biocentrism) or that of the *oikos* (ecocentrism); but as a result, deep ecology can only find or achieve this unity by way of rejecting humanity as such. The Chthulucene of the deep ecologists is a Chthulucene that has had one of its extremities amputated, seeking less to create new kin—like Haraway—than to rid itself of the cumbersome human kind. If the bio- or ecocentrism of deep ecology fails to adequately reject the dominant subject-object status (the Anthropocene), it's a result of deep ecology's inability to contest the redistribution that underpins it—namely, the call for a unity or the unconditional affirmation of the interconnection of everything with everything. The ontology that we are looking to develop must therefore find a way to leave a place for nature and, at the same time, *leave a place*, as such.

In order to be truly deep, to open up the depth of the field, ecology must become trenchant, fierce in a sense, focused—*sharp*: It must become a sharp ecology capable of simultaneously reappraising nature and distance. Since once everything is too interconnected, once nothing can truly claim individuation for itself alone, once existence—etymologically *being outside*—is impossible, once immanence prevents distance, nature ends up turning into its opposite: a nonnature.

HOMO NATURANS: MACHINISM AND NATURALISM

It would be incorrect for the eco-constructivists to rejoice from the critique we just put forth, since Spinozism is the repressed truth of eco-constructivism, its abhorred dialectical dark side [envers]. The truth of a world that, to refer again to Deleuze's expression that we cited earlier, *"does not make any distinction* [our emphasis] *at all between things that might be called natural and things that might be called artificial."* It's clear, Deleuze writes, in describing Spinoza's thought, "the plane of immanence or consistency"—i.e., Nature—is "always variable, and is constantly being altered, composed, and recomposed, by individuals and collectivities."[35] If, in fact, the expressions "plane of nature" or "plane of consistency" are made to be equivalent, if for Deleuze and Guattari there is "no distinction between man and nature: the human essence of nature and the natural essence of man become one within nature in the form of production or industry,"[36] if "everything is a machine,"[37] it's because in truth what Deleuze and Guattari describe is not an atemporal world but our contemporary society, where everything is recomposed, where this permanent recomposition is applied to quasi-individuals without any equivalent existential uniqueness, grasped within the contiguity of a hypercompressed substance without an outside, without any internal distance. This is the *Spinoza Society 2.0* that conjoins Spinozist substance with limitless hybridizations: nature and artifice become mixed together like the way in which a fly, a man, and a machine become horribly mixed together in the final sequence of the remake of the film *The Fly* (dir. David Cronenberg, 1986). In the acosmic world of the Spinoza-network, everything is merely a variation, a limited individuation at the heart of the "diversity" of a techno-natural continuum: existences without outsides.[38]

It's true, Guattari's ecosophy—being much darker than what we generally retain from it, and attentive (and far from any sort of Spinozism) to the various "death drives"[39]—was rather sensitive to the question of the finitude of "existential territories" as well as to the necessity for "reinventing alterity," as much in the "human domain" as in the "cosmic register."[40] But the pan-machinism that reigns throughout Guattarian ecosophy, its generalized animism,[41] will have only been able to confront, with great difficulty, the leveling of the world and the loss of alterity that flows from the confusion of the living with technical objects. In the end, Guattari will ask us to move away from a "backward-looking and irritated ecology resting on its laurels toward a futurist ecology that is completely mobilized toward creation."[42] Mobilization against irritation, future against the past, creation against defense: When everything is a machine, post-environmentalism spreads like a vapor trail of the virtual.

In such a fusional world, the greatest ruse of the official discourse of the Anthropocene is to present to us the human species under the auspices of *Homo naturans*. As we saw at the beginning of this chapter, nature, in the form of the natured, is from now on cut off from its naturing force. Matter waiting to be manipulated by anthropogenic force, nature is reduced to the level of an object. In front of it stands a conquering human subject who appears to have swallowed the naturing force whole. But we should be very clear here:

1. In no way does *Homo naturans*, as Spinoza had hoped for, bring together natured nature and naturing nature; rather, it reduces them: The natured is reduced to an object, while the naturing is reduced to vanishing for the benefit of *Homo naturans*.

2. We should, nevertheless, be somewhat suspicious of this appellation: *Homo naturans* is nothing more than the manner whereby the grand narrative of the Anthropocene designates its narrator; but behind the mask of *Homo naturans* stands the well-equipped *Homo faber* and his machines, *Homo faber* with a Wi-Fi connection to his "plane of consistency," *Homo faber* who appropriated for himself that which, during the Middle Ages, would have been attributed to God—techno-genesis having replaced divine creation. Of course, we could, along with Hannah Arendt, affirm the disappearance of *Homo faber*: Today, it's machines that would have automatically produced and dispossessed humans of their capacity for building.[43] In place of *Homo faber*, Günther Anders maintains the arrival of "*Homo creator*," who creates natural entities who, up until then, hadn't previously existed, creating physis by techne.[44] This being the case:

A) Anders is obliged to add that the human being is incapable of "creation ex nihilo" and that his true capacity is destructive.[45] Whether we qualify the human being as *naturans* or *creator*, in both cases it's technology that is euphemized. One must therefore be careful that the critique of technology does not participate in the technophilic dream of the naturalization of technics: *Homo faber* hasn't disappeared, but, rather automatically, humans dream of themselves as naturing.

B) This dream leaves one to believe that the source and origin of naturing would be found within *Homo*: And yet, nature imposed itself on humans, who have no capacity whatsoever for canceling out the existence of gravity or the speed of light! And this is perhaps the most profound fantasy that resides behind the appeal to a *Homo naturans*: a nature integrated within the human kind, mastered to the point where the technology used disappears into an absolute power of creation, capable not only of terraforming the planet and bodies but also of reshaping the laws of the universe . . .

From now on, we can see that it is necessary to make a clear distinction between:

—an absolute naturalism that maintains that nature is everywhere, including within the human being;

—and the fiction of *Homo naturans*, which maintains that humanity from now on has the natural power of being everywhere, of creating everywhere, to such an extent that technology as such becomes indiscernible from the naturing.

As we have seen throughout our study dedicated to Spinoza and to deep ecology, an absolute naturalism will be of no use to us in order to find some sort of exit strategy—the capacity for breaking away from the compact links and fusions of nature/technology and human/nonhuman. However, nothing prevents us from proposing a *naturalism of opposition* or a *naturalism of division*, not in order to uncover clearly defined borders between hierarchized beings but for adventurous passageways for singular beings. Cured of the sickness consistent with seeing naturally natured nature everywhere or seeing nature humanly natured, this naturalism of opposition would be able to nourish a trenchant ecology, capable of recognizing the anaturalism hiding under the mask of *Homo naturans*, and to reconsider and rethink machines in their alterity. Let's sketch the contours of this *sharp ecology*.

CHAPTER 11

THE REAL NATURE OF AN ECOLOGY OF SEPARATION

THE PRINCIPLE OF PRINCIPLES OF ECOLOGY

What actually programmed the definitive failure of the Spinozist cure? Whereas deep ecology's criticism of modern science is justifiable, why was it destined for failure when faced with staving off the constructivist paradigm? Because deep ecology fails to truly question the principle of principles of ecology. This principle is expressed quite simply in the following declaration: *Everything is connected*. This principle was perfectly formulated by Barry Commoner in 1971 in the book *The Closing Circle*. In his book, which had a resounding effect at the time of its publication, Barry Commoner describes what he names the four laws of ecology. All the other laws are derived from the first one: *Everything is connected to everything else* (a law that, we should note, Arne Næss cites in an essay devoted to the relationship between ecology and the philosophy of Spinoza).[1] Basing his thinking of relation on cybernetics, Commoner shows that feedback loops connect each part to the totality of the system (what affects one part affects the other). "Everything is connected" can also be expressed as: Everything is continuous, everything is continually in transformation, without disappearing (for Commoner, matter is indestructible). From which we arrive at the second law: *Everything must go somewhere*. This law affirms that there is no such thing as waste in nature, which—here comes the third law—is said to "know better than" human beings: *Nature knows best*. Commoner maintains that nature knows best because it has several billions of years of existence and research and development over humans, particularly with regard to the fact that human intervention cruelly lacks any sort of historical distance.[2] And finally, we arrive at the fourth and final law that stipulates "nothing is free": *There is no such thing as a free lunch*.

Because the global ecosystem is a connected whole, in which nothing can be gained or lost and which is not subject to over-all improvement, anything extracted from it by human effort must be replaced. The payment for this price can't be avoided; it can only be delayed. The present environmental crisis is warning that we have delayed nearly too long.[3]

Clearly and plainly elucidated, this fourth law of ecology expresses what appears to be the foundation of the current dominant theoretical ecology: Since everything is connected, to interrupt or damage a relation inevitably leads to damaging the specific ecosystem in which this relation is inscribed, as well as the ecosphere in general. As a result, above and beyond these analyses that date back to the 1970s, all of ecological thought rests on the principle of principles and its defense of relations at all costs. Born on this theoretical continent constituted by "the economy of nature," Humboldt's "cosmos," and the new "population" approach of Darwin and Wallace, ecological thought, Jean-Paul Deléage tells us, becomes "conscious of itself" at the moment when Haeckel invented the word *ecology* in 1866 and when new concepts began to flourish, such as the concept of the biosphere in 1875, the concept of biocenosis in 1877, and the concept of ecosystem in 1935.[4] At each one of these stages, whether in the Romantic approaches of the nineteenth century or the twentieth century eco-constructivist ways of thinking, as well as their *deep ecologist* enemies, Deléage's way of defining *ecology* is systematically verified: "a science of relation between living beings."[5] As such, John Muir, as a representative of an ecology linked to a certain American Romanticism, can declare that "when we try to pick out anything by itself, we find it hitched to everything else in the universe."[6] And, as far as deep ecology is concerned, Næss maintains that "we arrive, not at the things themselves, but at networks or fields of relations in which things participate and from which they cannot be isolated"; in this sense, "organisms and milieus are not two things—if a mouse were lifted into absolute vacuum, it would no longer be a mouse."[7] Resistant to any sort of Romanticism as much as to any kind of deep ecology, Latour nevertheless founds his ecological political thought on the concept of "attachments"; and Stengers speaks of "entanglements."[8] Today, we even find the so-called "new materialists" maintaining and reinforcing the principle of principles of ecology. In contrast to the constructivist and deconstructivist currents, the thinking of the "new materialists" has become very similar to ecological thinking and practice in its attempt today to give back a place to matter. Stacy Alaimo, one of the representatives of new materialism, in her book titled *Bodily Natures: Science, Environment, and the Material Self*, maintains that the substance of the human being is "inseparable" from its environment, and she provides the name "trans-corporality" to the process by which "the human is always intermeshed with a more-than-human world."[9]

Let's be clear: The principle of principles of ecology that connects—in spite of their differences—the romanticism of nature, eco-constructivism, deep ecology, and new materialism has a profound justification, and in no way is the goal of this present work to liquidate this principle, as certain strategies of thinking deemed to be object-oriented or a part of the current of speculative realism attempt:

—Infuriated by what seems to him as both an anthropocentric and relational excess, the philosopher Graham Harman claims that it is entirely possible to define the essence of an object by withdrawing it from any sort of relational network.[10] For Harman, in contrast to Næss, a mouse within an absolute vacuum would still be a mouse.

—For his part, Quentin Meillassoux is opposed to the thesis of "correlationism," according to which an object cannot be thought separate from the subject implied in the act consisting of thinking it. In complete contrast to this, Meillassoux maintains that an object can indeed be mathematically formulated without it being necessary to relate this equation to an intentionality, a consciousness, or a human-or-living-relation to this object. Against correlationism, Meillassoux's "speculative materialism" defends the possibility of a science without a subject.[11]

In our opinion, these two neo-Cartesian philosophies don't seem to be capable of avoiding nourishing the *refusal of relation* instituted with modern science—between the so-called "human" and the human's nonhuman "environment" (what Meillassoux defines as the "Great outdoors" devoid of life), between the glorious Industrial Revolution and the destruction of the conditions for the possibility of living, technologies, and "risks," etc. It's this refusal of relation that we have seen at work within geo-constructivism and that constitutes one of the points of friction with eco-constructivism and, more generally, with almost all of the thinking connected with ecology. From Aldo Leopold's ethics of the Earth to Michel Serres's "Natural Contract," passing by the "Gaia Hypothesis" supported by James Lovelock and Lynn Margulis and recently retranslated by Isabelle Stengers, from violent oppositions to deforestation that began at the beginning of the nineteenth century to the fights against GMOs and objectors to growth in the twenty-first century, thinking concerning ecology and the various engagements in environmental activism have all constituted themselves based on the idea that relation takes precedence and must be protected against what ravages it. Consequently, why should one contest this principle that appears as an excellent means for combatting Western anaturalism?

ON THE DIFFERENCE BETWEEN SEPARATION AND SPLITTING

For a very crucial reason that we will now set out to explain: If the battle against the great divide of nature/culture means to do away with any sort of separation, then this battle will do nothing but nourish a globalized anaturalism—which is what our study of Spinoza wanted to show. The displacement of terrain that must be effectuated consists in making it so that the end of the great divides is no longer the principal theoretical target of ecology, without wholeheartedly agreeing with the Cartesian and post-Cartesian denial of relation (object-oriented thinking and speculative realism). In order to enact this theoretical displacement, it is first necessary to understand that the denial or rejection of "humanist" relation, which grants an ontological position to the human being (thought, the awareness of death, etc.) that is refused to the other forms of life, or the denial of "Cartesian" relation, that excludes human consciousness from a world reduced to mathematizable and controllable matter, is not—contrary to what we might want to believe—a separation but a splitting [clivage]. Whereas separation articulates and recognizes differences, a split establishes an identity on a refusal of recognition. In other words, the denial or rejection of relation is founded on a split and not a separation. Let's explain the difference between the two terms:

1. A split is a radical rupture between two realities or two levels of reality. Indeed, we make use of a term originating from psychoanalysis: Freud speaks of a splitting (*Spaltung*) when part of the ego recognizes and accepts a reality that another part of the ego denies.[12] In certain cases of psychosis, the splitting is such that there is no communication—that is, no relation—between the two parts of the ego. The notion of splitting can also manifest itself through the presence of an event to which the subject experiences no affect whatsoever: a traumatic experience seemingly cut off from any affective dimension and, as a result, any cognitive dimension that is truly appropriate for this experience. To say "I know very well . . . but nevertheless" is a denial, in other words, a way of recognizing reality—"I know very well" that environmental problems are serious—at the same time rejecting this recognition—"but nevertheless," thanks to geoengineering, humans will be able to come away unscathed (or, at the least the humans residing in the Global North . . .). If denial (*Verneinung* is the term used by Freud) establishes a certain degree of reality, rejection (*Verwerfung*, which Lacan translates by foreclosure [*forclusion*]) is splitting brought to its conclusion: The rupture is absolute and reality is largely foreclosed (not recognized as such and subject to a kind of radical psychological blindness). As far as environmental questions are concerned, we could say that the larger the rejection of them is, the more inevitable the disasters. One example: The more

animals are simply considered as slabs of meat, as nonliving edible things—in other words, the more the experience of eating is split off from any cognitive and affective relation—the more the process of the industrial reduction of animality to the status of an amorphous material can develop. As we can now understand, anaturalism is founded on a massive splitting that enacts a dramatic rupture between human reality, on the one hand, and, on the other hand, a nonhuman reality to which every quality has been extracted. It's precisely on the basis of such a rupture that speciesism could be reconsidered: on the one hand, a privileged species (the human species), on the other, all the other species that pretty much count for (almost) nothing. In other words, a massive splitting is not a Great Divide of reality but a Great rejection of a large part of it.

2. But not every separation is a split[ting]. A splitting is a hermetically sealed border whose principal characteristic is not even being capable of being detected, of making itself invisible, inaudible, and imperceptible. When there is a splitting, we're not aware of it, the subject associates no affect whatsoever with the traumatic event. However, it is entirely possible to envision a separation that is detectable, fully experienced and lived, with emotion, and full of meaning. Such a separation, which we could also call an interval [écart], a deviation [écartement], or a distancing or spacing [espacement], is simply a distinction between two things, two beings, even two aspects of the personality. However, just because there are two distinct aspects of the personality doesn't necessarily lead, à la Hitchcock, to the *Psycho* syndrome. In the same way that it is completely possible to consider that human beings are separate from all other forms of animals without considering that this separation is of the order of an "abyss"—i.e., of an *integral rupture*. Where an integral rupture rejects the other through which it's supposed to constitute its relation of identity—we could take Heidegger's example, the idea that the animal is "poor" of the world, whereas the human is *world-forming*[13]—in complete contrast, separation is the recognition of the other. Since to recognize, for a human subject, even if he or she is simply existentially separated, refers back to recognizing a dependence (and not a "pressure") vis-à-vis the others with which he or she shares the same world. No human being is self-sufficient—I am connected to others because I am separate, because I bear within me alterity—the alterity of the language I have learned, the inner alterity of everything that has been transmitted to me. To take back up one of Saint Augustine's equations: Alterity is "more interior than my own intimacy (*interior intimo meo*)."[14]

From then on, a fundamental problem for ecology shows itself in the light of day that we can formulate in the following way: To take as one's target any divide—in other words, any separation—can only have an effect of destroying relations. In making an apology for the One, of an absolute substance, of continuity, of hybridization and attachments, or human-nature identity, the eco-constructivist and naturalist thought of ecology (of the Spinozist, *deep ecologist*, Deleuzoguattarian type) confronts a wall of confusion—a confusion stemming from the end of separations. Ecological thinking doesn't even risk entering into the "night where all cows are black," to borrow Hegel's expression when he mocked Schelling's philosophy of the absolute (a philosophy that, according to Hegel, cancels out all differences), it doesn't even know that it has entered into an indistinct night, of no longer even knowing whether it's a cow or noncow, human or nonhuman, prions or dauphins. By fighting against separations, by fusing together strata of reality, an ecological thinking will do nothing but undertake an anaturalism through a radical colonization of everything that is nonhuman by the human, albeit through the painted-over face of *Homo naturans*. Since, as we have seen over and over again, the eco-constructivists have rejoiced in singing the end of nature but not that of technology, by doing away with the great divides of nature/culture, nature/technology, eco-constructivism has let technology and culture unfold within a nature that has been annihilated in advance. Conclusion: We must destroy splittings but not separations! This will be one of the passwords of an ecology of separation.

ESCAPING THE TRANSCENDENTAL NARCISSISM OF *HOMO NATURANS*

What sort of nature could correspond to this ecology of separation, which refuses both splitting and fusion? This nature must simultaneously be in relation with human consciousness, thought, yet without being reduced to it. A reduction of nature to thought in effect leads directly to what we could call a transcendental narcissism. The latter consists of positing nature as a) that which exists only as an object of thought, of philosophical speculation, and b) that which is therefore shaped, modified, and reshaped according to this thought when this thought becomes equipped—technologically armed, anthropogenically efficient. Defined as such, from then on, nature would become ideally and effectively the product of our thought and our actions. This is precisely the narcissistic trap that Werner Heisenberg, one of the founders of quantum mechanics, described in the 1950s:

1. With quantum theory, far from allowing us to gain access to a reality uncorrelated to any human subject, "the new mathematical formulae no longer describe nature itself but *our knowledge* of nature": Observation disturbs reality at the scale of elementary particles, what we grasp is ourselves within the experience to which we subject subatomic nature, because it's ourselves that we have projected in advance within and by way of the theory.[15] It's no

longer nature that is written in mathematical language, as Galileo maintains, but, first and foremost, humanity.

2. As a result of our extraordinary technological capacities, this projection is not merely theoretical or imaginary, it is material and consists of transforming the world in such a way that, in the world, "man confronts himself alone."[16] From this anthropocenic situation that is somewhat schizophrenic, Heisenberg provides the following striking image:

> In what appears to be its unlimited development of material powers, humanity finds itself in the position of a captain whose ship has been built so strongly of steel and iron that the magnetic needle of his compass no longer responds to anything but the iron structures of the ship; it no longer points north.[17]

Against this transcendental narcissism—this replacement of nature by humanity—and its technological translation, it becomes necessary to posit a certain kind of realism—i.e., a separation of nature, but a separation that would not lead to a splitting. This is precisely what Whitehead's philosophy of nature proposes, a philosophy we will follow with great attention. For this philosopher, there is nothing worse than considering nature first and foremost as a matter plunged into time and space, as something natured; for Whitehead, it's completely the opposite: Nature is of the order of naturing nature, it is a passing, an advance, a process, a passage, activity, it is a collection of events before being facts.[18] That said, Whitehead's philosophy perfectly resists the naturalist temptation: Nature is not everything, nor is it the Whole; it is a cosmic dimension of reality that the general cosmology of *Process and Reality*—Whitehead's magnum opus—takes as its task to think. The progressive advance of nature is the way in which time itself advances, the time of pure events, which is built through objects that endure. Contrary to the approach that makes nature bifurcate between, on the one hand, a substance in itself, pure scientific entities, and, on the other, attributes that would be added on to these entities by consciousness, Whitehead offers us a *theory of internal relations* that we will adopt here in the following manner:

1. Nature is a field of intrinsic relations that reconnects the event to perceptions. In this sense, substances and attributes are, respectively, degenerated perceptions and events (i.e., abstractly cut off from the advances nature makes toward us).

2. We must clearly understand the consequences of this approach and the danger it neutralizes: If nature is a field of *extrinsic* relations, considered as added from the outside to entities, it would then be possible to replace these relations with other relations, to build them—to rebuild them. This is the

ultimate fantasy for the humankind of the Anthropocene, *Homo naturans*, in his cybernetic outfit. By considering nature in measurable quantitative terms (for sale, exchangeable), we are witnessing the dispossession of *naturans* for the benefit of *Homo*: What had been posited—deposited, composed—by the spirit can be taken back up by the mind—i.e., recuperated and remade. Against a generalized mechanism, and in aligning ourselves on this point along with romantic thought, we must think the idea of nature in organicist and qualitative terms: Nature is what is organized through relations—in other words, it is self-organizing.[19]

Neither mechanist nor vitalist, Whitehead offers us a philosophy that avoids turning nature into a cosmophagic totality or turning the world into a naturophagic machine. In fact, the self-formation of nature implies a separation vis-à-vis consciousness. And here we have one of the central arguments of our thesis: The disclosure of nature to consciousness doesn't do away with nature's opaque feature, the fact that nature is inevitably closed off to the human mind. As a consequence, when I perceive nature, there is *more* than what I can take in, that is, there is a definitive nonego; Narcissus's mirror wasn't created by Narcissus: The lake into which Narcissus gazes at his reflection is one of the expressions for the progressive advances of nature, the progression of the Earth throughout millennia, well before Narcissus was able to add his gaze to it. Nature truly serves as a limit, not in the usual sense of nature as limited resource but as an ontological limit: Everything is not, *cannot be*, the effect of an anthropomorphic projection. However, our interpretation appears to at least partially run counter to what Whitehead claims: "This closure of nature does not carry with it any metaphysical doctrine of the disjunction of nature and the mind."[20] It's true: Whitehead does not want to establish a dualism—a disjunction—of the mind and nature, as if the human mind contemplated nature from the outside! But it is also imperative to recognize that nature discovers itself to be closed off from the inside of perception. Not only is Narcissus's mirror not Narcissus—it wasn't originally created by him—but the mirror is, at least partially, opaque.

THE COUNTERPRINCIPLE OF SEPARATION AND REAL NATURE

Thanks to Whitehead, we seem to have found a way to escape Spinozism (its acosmism, its difficulty in providing a place for nature and for existing entities), of avoiding naturalism (the excess of nature), without having to return to a Cartesian position. From now on, we are aware of the fact that the principle of the principles of ecology is inseparable from a counterprinciple of separation. It is possible to escape from the immanent network that founds the principle of principles of ecology without invoking acosmic objects as is done by the object-oriented ontologies. Not everything is nature. But nature makes it so that everything is not human.

What we call *real nature* is neither the totality of the world nor an amorphous material but that which, without being an absolute subject or a limited object, objects to the consciousness granted to it. It is therefore, neither nothing nor everything. It is naturing without dissolving that which isn't it. It is natured without reducing itself to a compact object. Far from serving as the great connector, as the mystical ocean where each drop is called to joyously dissolve into the others, real nature can from now on act as an instance of separation within a world that has extended the principle of principles of ecology to technological mediations, to the *world wide web*, to the proliferation of the chatty objects that make up the internet of things. When the global ecosystem becomes a limitless hybrid techno-ecosystem, real nature serves the function of being a kind of circuit breaker. Therefore, an ecology of separation doesn't have to separate us from nature in order for us to dominate nature, since it makes use of nature in order to separate us from our techno-economic automatisms. Nature is no longer cut in two, between the naturing and the natured; nature becomes cutting, apt for expressing the counterprinciple of separation for toppling the empire of the principle of principles of ecology.

A question nevertheless remains: How can we explain the opacity of real nature? How can we account for this seclusion? Are we dealing with a power of naturing nature? Or, completely the opposite: a limit that affects nature at its core? And what if the lack of immanence of nature was the definitive beneficial effect of that which is counterproductive or counterprocessual within it?

CHAPTER 12

DENATURING NATURE

ANTIPRODUCTION

To begin the third part of our present work, we have focused on the division that affects nature between naturing nature and natured nature. But we also clearly see the impasse to which this division leads us: 1) Either we maintain this division and we find ourselves falling over to the side of a geo-constructivist Cartesianism that props itself up on a relation of splitting and a definitive rejection of nature; 2) or we reject this division and find ourselves with the hyperconnectionist option, be it in the form of that found within *deep ecology* (which is profoundly Spinozist) or that of eco-constructivism (Spinoza in the age of the *world wide web* and proliferating machines). How can we escape from this impasse? How can we affirm nature as advance, as passage, as process, and at the same time maintain a form of separation? Whitehead led us very close to the solution, but we have no idea why nature is opaque—that is, why it is that nature reveals itself (including in the form of Galileo's "book of nature") without giving itself to us (unlike the nature sculpted by Louis-Ernest Barrias in the form of a prostitute). There must be something in nature that pits itself against nature, a tendency that is at once natural and antinatural, without, however, constituting a transcendent entity (God), that allows for the advance or forward movement of nature at the same time as it allows for the possibility of not advancing or progressing, of not developing. Something that allows nature to continue and, at the same time, not continue . . .

Hic Rodhus, hic salta: We propose to call this part of nature *natura denaturans*—denaturing nature—that which, from within nature, as nature, is opposed as much to the productivity of nature as it is opposed to nature as a simple product. So as to conceptually establish this idea, we will make use of the way that Schelling, in complete contrast to Spinoza, subverted the traditional break between *natura naturans*, defined as subject of limitless productivity, and *natura naturata*, defined as the stable product of nature as an object.[1] Indeed, Schelling duly noted that a product is never absolutely stationary, it is always in a transition toward another state, in other words, it is always passing within an infinite productivity, it is nothing more than its temporary repository. Temporary? Well, then how do we explain the existence of something that is finite, stable, this object, this world! "How can the absolute escape from itself and oppose a world to itself?," Schelling wondered in 1795 in his *Philosophical Letters*.[2] The solution he proposes, and that we will put forth in our own present attempt, is the following: Infinite productivity must counter itself, it must put the brakes on *itself*, it must slow *itself* down. Nature can't only be forward advancement, *nature must also lag behind*, it must also constitute its own "constraint."[3] This lag can be defined as an "antiproductive" tendency without which all productivity would be vain: There would only be the absolute, the infinite, endless movement, and as a result, nothing that would be finite, no object whatsoever. This internal difference between production and antiproduction is the reason why nature is split.[4] But we see to what extent we are now very far away from the divide that frightened Spinoza so much; the divide between naturing/natured: *for the true division does not go between naturing and natured but between naturing and denaturing*—i.e., between productivity and antiproductivity. In this sense, and from then on, as distant from the fanatics who consider nature as an infinite subject as much as the object fetishists,

> in Nature, neither pure productivity nor pure product can ever exist. The former is the negation of all product. The latter, the negation of all productivity.[5]

To better understand this idea, we should remember that Deleuze and Guattari also make use of the concept of "antiproduction," but in negative terms and according to three distinct modalities that we find useful to briefly describe:

1. A total rejection of the schizophrenic kind (the refusal of everything that was other than the "body without organs"—i.e., the body withdrawn from any sort of organization, order, hierarchy, and functional specificity, that the schizophrenic, according to Deleuze and Guattari, constructs for himself).

2. A lack artificially produced by the State right where a permanent abundance should reign. This lack is the effect of a body exempting itself from the circuit of production, passing itself off as an "ungendered nonproductive attitude": In a paradigmatic twist, it's the body of a despot that "falls back on all production," "thereby appropriating for itself all surplus production and arrogating to itself both the whole and the parts of the process, which now seem to emanate from it as a quasi-cause."[6] In a certain manner, this so-called

ungendered body is the political equivalent of the schizophrenic's "body without organs."

3. Within the capitalist regime, an omnipresent death instinct, immanent to production:

> because, by means of the immanence of the decoding, antiproduction has spread throughout all of production, instead of remaining localized in the system, and has freed a fantastic death instinct that permeates and crushes desire.[7]

In contrast to Schelling, Deleuze and Guattari—whose theory fuses Spinoza and the positions of the young Marx of the *Manuscripts of 1844*—do not seem to apprehend that far from being something terrible to conjure, antiproduction is what allows for a world to exist. The division that affects real nature does not imply some sort of destructivity that would come from the outside, but it is the constraint, *the impediment that allows* for productivity to endure.[8] In the same fashion, antiproduction doesn't add itself to production: In order for something to deploy itself, the freeing creative movement of nature must be constrained; in order for there to be an advance, there must be a lag. *Lag and advance are the two sides of real nature.* Of course, and Deleuze and Guattari are correct when they argue that antiproduction can become autonomous and turn into a death instinct, into destructivity: It then goes from the status of an internal negativity—the necessary constraint for there to be the finite, the natured—to an external negativity that leads to one of the multiple deaths of nature we studied at the end of the second part of this present work. In this sense, the environmental discourse that largely insists on the necessary and happy death of nature does nothing but reinforce destructivity instead of taking care of the antiproduction without which no existence, no world, is possible.

WITHDRAWAL OF NATURE AND DARK ROMANTICISM

The *natura denaturans* is of the order of an antiproduction. But this antiproduction seems to follow a production, and we could imagine a Deleuzian or a Spinozist assuring us that the denaturing character of nature is still the affirmation of its creative force. Antiproduction is still production! Consequently, what we must still be able to show is that antiproduction precedes production. That the manifestation of natural things presupposes a nonmanifestation—a dissimulation, a withdrawal, an initial seclusion. However, far from being original or rare, this idea has expressed itself throughout history in the form of an infinite commentary on one of Heraclitus's fragments that has long irrigated the philosophy of nature: *physis kryptesthai philei*, "nature loves to hide." According to Pierre Hadot, this fragment deals with nature's strange habit of "disappearing" whereas, according to its primary meaning, it should only be a process that consists of "appearing."[9] *Whereas nature should only contain the naturing, nature nevertheless tends to conceal another tendency.* This tendency for the disappearance of nature does not simply mean that everything must die after having lived, that everything must disappear after having appeared, but that nature, before naturing, creating, appearing, begins by disappearing, by veiling itself, by hiding, before then flourishing. Here we find ourselves quite far from the thought that turns nature into an identifiable object, a natural resource. But it's not about thinking of nature in terms of a source, if we mean by this term that which suddenly springs forth (nature as principle of growth and transformation of which Aristotle speaks).

The crypt of nature—an expression that the word *kryptesthai* invites us to use—can be explained through some of Heidegger's terms such as the "withdrawal" of being that precedes the way in which being "confers itself to us."[10] Being, Heidegger writes, "withdraws itself, through showing itself as being as such": Remaining "lacking," "reserving" for itself what would be a complete "unconcealment," Being is the "promise" of future beings, of future manifestations.[11] As Marlène Zarader shows, this inaugural withdrawal is strangely close to the doctrine of Tzimtzum elaborated by Luria, a fourteenth century Kabbalist: If God is everywhere, the lone possibility for there being a world is that God "contracts," "withdrawing into himself, to put it that way."[12] God is not hidden within the world but within himself, so as to let the world be. The revelation *follows* from the withdrawal: "Whereas he [God] withdraws with himself, the Infinite leaves behind him a free space. In this manner, he creates a lack, similar to darkness."[13] And yet, well before the arrival of Heidegger, and in perfect harmony with the Kabbalist idea of contraction, Schelling maintains in *The Ages of the World* (in 1815, that is, sixteen years before his *First Outline of a System of the Philosophy of Nature*, which we have explored previously) that everything begins, the world as much as each being, by way of a "contraction," a "negation": "But all beginning is founded on that which is not."[14] Indeed, growth and expansion are only possible by way of starting with their opposites and not by way of starting from a neutral state where there would be neither contraction nor expansion—since nothing then would allow us to understand the passage from the neutral to the positive—in other words, the manifestation of things.[15] For Schelling, the originary negation that precedes the positive presence of entities in the world never disappears, the original negation is "still the wet-nurse and mother of the entire world that is visible to us";[16] night always remains.

Night is, in fact, the best possible image for taking into account this negation, this first negation and its cryptic effect. Whitehead, of course, spoke of the "Romantic reaction" as a rejection of the industrial world that had become incapable of recognizing the "value" of everything that "persists" as a limited, singular entity incapable of entering into the capitalist circuit of exchange without withering and perishing. But the singularity, the capacity of each being to not be engulfed within the "non-being of indetermination,"[17] paradoxically calls for a *nocturnal share* that is the trace of the contraction by which every individuation begins. Beyond Whitehead, it's Schelling's *dark Romanticism* that is conceptually required for grasping the way in which each being becomes singular: It's not *conatus*—i.e., "the effort for persisting in one's being" (Spinoza)—that allows us to understand the way in which each being is unique, simultaneously taking the world within itself and also flourishing; it's the effort expended by the nonbeing to persist after its initial *negative explosion.*[18]

This dark Romanticism is certainly disconcerting, and the night of the world that we choose here as a metaphor of the denaturing nature was firstly a Romantic topos initiated by the fear of God. With this topos, we can detect the metaphor of the "empty eye" of God evoked by Jean Paul Richter,[19] this precursor of Romanticism, which will be repeated notably by—among others—Gérard de Nerval in his poem, "Christ on the Mount of Olives":

Seeking the eye of God, I've only seen a pit,
Huge, bottomless, black from whence eternal night
Streams out over the world and ever deepens.[20]

From Novalis's *Hymns to the Night* to Hugo's "What the Mouth from the Shadow Says" in passing by Keats's "tender night" ("Ode to a Nightingale"), Lord Byron's "Darkness," and the "black sun of melancholia" of Nerval (*El Desdichado*), the poets and philosophers who participated in the rise of the Romantic movement (Schelling) or who without question were inheritors of it (Hegel) understood that the death of God—the collapse of transcendence—brought to the light of day a nullifying force, a negative and nocturnal force, anterior to the divine light.[21] It's this frightening night of the world that Hegel uncovered within each human being:

> Man is this Night, this *empty nothing* [our emphasis] that, in its simplicity, contains everything: an *unending wealth of illusions and images* [our emphasis] which he remains unaware of—or which no longer exist. It is this Night, Nature's interior, that exists here—pure self—in phantasmagorical imagery, where it is night everywhere . . . where, here, shoots a bloody head and, there, suddenly, another white shape—only to disappear all the same.

We see this Night whenever we look into another's eye—into a night that becomes utterly terrifying—wherein, truly, we find the Night of the World suspended.[22]

An "empty nothing" and "an unending wealth of illusions and images": Is this not the dialectic at work in the paintings of Lucio Fontana that make up his series from the 1960s called *Spatial Concept*? Like a delayed reply to the *heavenly trembling* of the Romantic movement, these paintings are monochrome surfaces that have undergone several lacerations: each one of these interior openings uncovers, on the backside of the red paint, an impenetrable black gauze, like a night forever withdrawn. These paintings are not merely objects (products) or productive subjects (a productivity) but spaces slashed by an obscure absence that opens the paintings up to more than themselves—unto a distant and withdrawn and archaic dimension, *wild like the night of time*.

WILD IS THE PRESERVATION OF THE WORLD

What should we understand by the term *wild*? A category thanks to which we will bring to bear our research dedicated to the denaturing of nature within questions of ecology. It is, of course, necessary to protect what is wild in the name of defending "diversity," but what is wild is not simply one element among others, a vast terrain or an industrial wasteland: It constitutes the foundational ontological condition for all (bio)diversity. It is not simply that which *appears* as being outside or beyond domestication, rebelling against civilization, a bundle of chaos that escapes our mastery, but first and foremost, it is an indication of the remarkable denaturing *potentiality* of nature, and in this sense, it is the always preserved promise of the ever-present form of another society to come.

In order to set up our philosophical proposition, let's return to Thoreau's famous saying: "In Wildness is the Preservation of the World." As Ursula K. Heise notes, this declaration has been used many times to justify erroneous conceptions of the wilderness as a "pure" wild state of nature.[23] Furthermore, the idea itself of the Anthropocene seems to indicate the end of any conception of a region or area we could consider as wild. What's worse, to speak of wild space as such would be to confirm those well-known dangerous dichotomies (between what is wild and what is civilized, between nature and culture, etc.). Conversely, to speak of artificialization would be to—wrongfully—believe that technique is distinct from nature. Consequently, why not simply abandon the category of wild? For the very reason that the "great cementing" (Jade Lindgaard) continues. Over the past thirty years, there has been more than a

70 percent increase in the number of artificial surfaces constructed in France, as if every ten years another region of the countryside has disappeared under concrete.[24] Consequently, "There is nothing left that is wild" would no longer be a descriptive declaration but the gray underbelly of the Anthropocene: the declaration of a new wave of artificialization. In this light, preserving the category of the wild would perhaps be a way of granting oneself the means by which to critique the politics of a general cementing of the landscape. Remember that by the word *Wildness*, Thoreau didn't simply mean a place (which the term *wilderness* inherently implies) but rather a *dimension of time*: the possibility of a future symbolized by the West—the West as an object of a so-called "conquest" during the nineteenth century and that for Thoreau became the eternal metaphor for symbolizing the not-yet-conquered. The problem is that the future, in the manner in which Thoreau envisioned it, is difficult to discern from the repetition of the past when colonial powers conquered the "new world." If we must reject this schema of a Conquering West, of the West in general [L'Occident], it is also possible for us to grasp the way in which Thoreau strived to uncover an *inner outside*, counter to a certain kind of civilization that over-civilizes itself, thereby over-cementing itself and destroying any rapport with exteriority, any rapport with an *inner outside* from which the world could begin to give birth to itself anew. It is indeed these kinds of inner outsides that are lacking today within the Anthropocene and its effects of homogenization, its all-encompassing grip upon everything toward its reformatting.

It is in this light that today we can begin to reconsider what the word *wild* means: not a mythified space, not the West, and therefore, not the *wilderness* as a presumed symbol of the source of a nation; but it also wouldn't mean the rejection of the remarkable in favor of the banal, of a blade of grass growing within the cracks of a piece of asphalt and the isolated flowers found in a city garden still free for the moment from being completely overrun by the Anthropocene. It's not enough to simply say that *wildness* as opposed to the *wilderness* (which is always linked to a specific place, such as a national park) can be found everywhere, at the top of a mountain or in my backyard: If that were the case, why would we even worry about protecting and preserving any area in the world? Why would we bother fighting against the transformation of national parks into areas that would be exploited for their large reservoirs of energy? To put it another way, what matters here is saying and showing how the state of the wild can be *rediscovered*. Yes, everything *can* be wild—behavior, sexuality, dreams, and, in this sense, as Gary Snyder would put it, wildness "*is* the world."[25] Solitude, alterity, and wonder can indeed be experienced anywhere on the planet; but this experience must be supported by an appeal for air or, rather, an *appeal for being*, a valorization of these experiences as something that is amazing—singular and memorable. In this light, the position held by William Cronon is crystal clear: It is important to recognize that there are things "that develop without our intervention," and the advantage of supposed wilderness areas is that they are more propitious than others for making us experience that "a tree has its own reasons for being, quite apart from us."[26] The problem, Cronon tells us, is that we tend to forget the otherness of the tree *we* have planted in our gardens. Cronon also doesn't call for us to eliminate the wilderness, he even thinks that it can be an aesthetic site, where the otherness that we tend to forget in regard to nature would be able to be revived, if and only if this site of wilderness is no longer thought of as absolutely and truly *split* from urban sites, from civilized spaces, but is considered as *integrated* with them in a "full continuum of the natural landscape."[27] Once this community is achieved, once the useless outmoded splits have been overcome, the unique differences can appear as such. In this sense, Cronon calls for what we could call a *democratization of the wilderness*.

Let's translate these analyses into ontological terms. Instead of considering the wild as an element of biodiversity, a form of nature among others—according to the schema of a *multinaturalism of diversity*—we must think of it as a condition for the possibility of difference, the *transcendental differential* [différenciant] allowing for something to become separate and individuated. As the philosopher Mark Woods writes, "what is wild exists as an expression of autonomy of that which is more-than-human, beyond anthropogenic controls";[28] but the wildness of individual autonomy is merely derived from a wild state that separates and articulates individuals among themselves. As transcendental differential, the wild is a spacing that is neither object nor material, nor process, but that which allows for movement; it is the condition for all transformation. In this sense, in order to return to Thoreau's famous saying, the wild is indeed a call toward that which does not yet exist—a way of living a form of society, a nature that would be less ravaged. A call to "wilding" (as Serge Moscovici would put it), so that the "forces of beginning" may emerge: It's not a return "to" but a return "of" nature:[29]

To render life wild is to de-massify it, to aerate the space and let it breathe.[30]

The idea of nature being outlined here is not one signaling toward some elsewhere outside of the world but toward the possibility of separating from the world in order to finally form a new alliance with it. The return of nature is an instance of disconnection, this return allows one to turn away from the world in order to reinhabit it in a new way. It is in this sense that the wild is the promise of a disconnection from and a reformation of civilization, this wild precedes everything that exists and projects us toward what could, eventually, exist. It is the somber precursor of our unsatisfied desires, the night of unfinished natures, the still-living trace of denaturing making its return within the very heart of that which would like to do away with nature.

WHAT IS THE UNCONSTRUCTABLE?

The wild is of the order of the unconstructable—in what sense should we understand this term? First of all, the unconstructable aspect of nature does not refer to something that is not constructed: "Nature is the primordial—that is, the nonconstructed, the noninstituted," Maurice Merleau-Ponty claims.[31] And yet, we know that what is nonconstructed today can quickly become the constructed of tomorrow! Thus conceptualized, so-called primordial nature offers no creative resistance whatsoever to the contemporary geo-constructivist drive—what today has always seemed natural (such as genes) will no doubt end up being produced by human beings tomorrow. As a result, we must then consider as unconstructable that which escapes all construction—whether past, present, or future—and, as a consequence, *precedes the primordial*.

However, we are not seeking to locate some sort of unassailable, impossibly manipulable terrestrial material that would, by some kind of miracle, resist an external process of destruction. The unconstructable does not escape destruction because it is indestructible, or because it is fleeing from death, but because destruction itself requires the unconstructable. Every action—whether constructive or destructive—requires a contraction, a subtraction, an antiproduction that precedes it as its dark side or counterlining *(revers)*.[32] The unconstructable is the inaccessible transcendental of production, which we will call its *transcendental dark side or counterlining*. Without going into an analysis of the concept of the transcendental, we will simply note that for Kant this term designates the a priori of all knowledge, an a priori independent of all experience.[33] However, unlike Kant, we will conceive of this transcendental dark side not as epistemological, linked to knowledge,[34] but as ontological, related to the existence of beings. Moreover, the transcendental dark side is not merely an a priori, a condition of possibility—for example, the wild to the extent that it makes diversity possible—but it is also a condition of impossibility; it gestures toward something unavailable: We cannot access the wild, nor can we grasp or reconstruct it. The desire to grasp it, in order to reconstruct or destroy it, involves movement in the direction opposite of grasping it, a countergrasping that is utterly unconstructable.

We could say that this very mysterious transcendental dark side or counterlining, with its negativity, is far removed from the real nature that Whitehead taught us to consider. In what follows, we will demonstrate how the unconstructable has always been connected to the real nature of the Earth throughout the continual transformation of the Earth. The unconstructable is not an event that took place just once, at the origin of the world and of things; it has been present throughout the entire history of the Earth, connecting its past with its most distant future: The transcendental dark side or counterlining is the other side of the concrete history of the Earth.

CHAPTER 13

THE UNCONSTRUCTABLE EARTH

THE EARTH AS FULL BODY: NEO-ORGANICISTS VS. GEO-CONSTRUCTIVISTS

For the geo-constructivists, the Earth is *nothing more* than an empty box, a perfectly modifiable piece of machinery, anatural through and through. For this very reason, all the opponents of this maniacal stage of the Anthropocene, and its various practices that involve wanting to control the ecosphere, appeal to the image of a living Earth, one that is whole, organic—*more* than a simple piece of machinery—but how do we define this *more*, this excess? This is the question that organicist and neo-organicist theories of the Earth seek to answer. Whether their answers are accurate remains to be seen.

Carolyn Merchant sees each part of an ecosystem as contributing in an essential way to the "stability" and "healthy functioning of the whole."[1] This holistic representation of nature can be found in the work of the founders of American environmental and ecological thought (Muir and Leopold), who emphasize networks of interrelations and believe that humans belong to a biotic community. This holistic view of nature is also at the heart of the organic geography of the explorer and scientist Alexander von Humboldt, who saw nature as "one great whole animated by the breath of life," from "the vault of heaven, studded with nebulae and stars" to the "rich vegetable mantle that covers the soil in the climate of palms."[2] Also in the nineteenth century, the geographer Carl Ritter wrote that his aim was "to treat geography as a systematic whole, and to show that in it, as everywhere else, the part can only be known from a conception of the living whole."[3] In 1896, Paul Vidal de la Blache, founder of modern French geography, described the Earth as an "organism" in which nothing "exists in isolation."[4] Whether consciously or not, contemporary minoritarian discourses draw on these earlier sources, which refuse mechanical reductionism and see the Earth as a body full of life and irreducible to physicochemical components. While the geo-constructivists want to cure us of our post-Copernican nihilism by intensifying technological mastery of the Earth, the neo-organicists emphasize the fact that the Earth is a heavenly body that is alive, not just adrift.

However, a very careful examination of James Lovelock's theses should lead us to question the simple opposition between geo-constructivists and neo-organicists. Lovelock, who along with Lynn Margulis proposed the Gaia hypothesis, compares the Earth to a living being, a "vast organism"—while at the same time he is a strong proponent of nuclear energy and fracking who describes environmentalists as fanatics and ideologues![5] Is it possible to be both an organicist and an antienvironmentalist? At first glance, Lovelock may appear to be following in the footsteps of Ritter and von Humboldt (von Humboldt's thought has influenced Crutzen and Schwägerl[6]); but, in fact, Lovelock reread and rewrote the organicist proposition through the distorted lens of cybernetics. Lovelock defines Gaia as:

> [a] totality constituting the feedback or cybernetic system which seeks an optimal physical and chemical environment for life on this planet. The maintenance of relatively constant conditions by active control may be conveniently described by the term "homeostasis."[7]

And yet, one of the fundamental objectives of the first cybernetics was precisely to erase the difference between living beings and communication machines—both of which, according to Norbert Wiener, the founder of cybernetics, sought to "control entropy through feedback."[8] If, as Clive Hamilton has shown, Lovelock has moved over the years from a rather vitalist conception of the Gaia hypothesis to a patently mechanistic view, it's because this transition was actually possible from the very beginning. From his very first hypotheses, Lovelock equated machines with living beings and mechanism with vitalism. So it's not that surprising that in his most recent works Lovelock ends up affirming that the Earth is living because it is capable of self-regulation, much like a simple machine equipped with multiple thermostats.[9]

His intentions were nevertheless good. In 1979, when Lovelock proposed his Gaia hypothesis, he clearly did so in opposition to the nihilistic version of Fuller's metaphor of the Earth as a "spaceship" and in opposition to its Cartesian foundations. For Lovelock Gaia is:

> [an] alternative to that pessimistic view that sees nature as primitive force to be subdued and conquered. It is also an alternative to that equally depressing picture of our planet as a demented spaceship, forever travelling, driverless and purposeless, around an inner circle of the sun.[10]

But clearly it is not enough to consider the Earth as an *auto*-regulated system so as to resist the temptation of a future *hetero*-direction—that is, the temptation to replace the *auto*, the *self* of the Earth, with the human—in other words, to change the pilot. Such a temptation relies on the belief that a "living" system steers *itself* in the same way a maritime pilot steers a ship. This comparison signals the inadequacy of a certain kind of organicism to counter—in a theoretically effective way—the prevailing anaturalism and its technological fantasies. The Earth may be a full body, but that does not prevent its interior from being—literally and figuratively—enucleated, modified, and artificialized by the science of the doctors of terraforming. The main problem, then, is clearly not the relation between an indivisible natural whole and its parts (organicism); the problem is knowing what kind of plenitude would be able to resist the stranglehold of geo-technology. Above and beyond its auto-regulation, we must consider the so-called living system, beginning with its auto-creation: the generative capacity that makes it intrinsically able to escape the powers of human control.

It is precisely this generative capacity that Margulis and Lovelock struggled to understand. When she defines the Earth as a partially living Being, Margulis is suggesting that humans will never be able to put "an end to nature" and that tropical forests "will continue their cacophonies and harmonies long after we are gone."[11] Her aim is not to acknowledge the Earth's generative capacity but to identify the natural—the ordinary, regular, and ultimately unremarkable—supremacy of the Earth over human beings. Similarly, for Lovelock, "the very concept of pollution is anthropocentric" in that it conflates the harm pollution does to humans with the Earth's auto-regulation and auto-adaptation. From the viewpoint of Gaia, the function of climate change—it's only function—is to maintain homeostasis without necessarily sustaining human life.[12] And yet the Earth's generative capacity—its wild capacity—becomes apparent when we are able to glimpse its unpredictable behavior, as climate expert Wallace Broecker and the journalist Robert Kunzig explain: "Every now and then, it seems, nature has decided to give a good swift kick to the climate beast. And the beast has responded as beasts will—violently and a little unpredictably."[13]

Building on the same metaphor, Clive Hamilton criticizes geoengineers for not recognizing that the Earth is "an uncooperative beast" that will not respond to atmospheric manipulations as we might prefer.[14] The characterization of the Earth as living or quasi-living is not done so as to accentuate the organic interconnection between everything on Earth but to take into consideration the surplus that results from any project of technological dominance. It's precisely this surplus that makes the Earth into a wholly full body: not a body filled with matter or organs, but with potentialities that no system—whether technical or living, artificial or organic—is able to contain.

THE INACCESSIBLE SIDE OF THE EARTH AND THE HUMAN "CONDITION"

In taking into account the discussion in the previous section, the following idea emerges: Geo-constructivism and neo-organicism struggle to offer an adequate image of the Earth and its existential surplus. Is this not ultimately because it's as if both these approaches seem to secretly be in a shared agreement by accepting the representation provided by Apollo 17 in 1972: the Earth as a "marble"—either empty and lifeless (as the geo-constructivists maintain) or filled with life (the neo-organicist response)? To be very clear: Our goal is not to place these two representations on the same plane but to entertain the possibility that perhaps neo-organicism is not capable of completely making the Earth full *enough* (i.e., sufficiently opaque, solid, and inaccessible). In order to make the Earth opaque enough to resist the technological dominance of the hypermoderns, we must emphasize its unconstructable part, which is not in opposition to its living part but precedes it, supports it, and can be aligned with it. This nocturnal, unobjective, and asubjective part is that which withdraws from human dominance and is subsequently established as the unattainable condition of humanity.

This inaccessible side of humanity, this imperceptible counterlining, is neither the Earth-object nor the Earth-subject:

1. The geo-constructivist Earth-object illustrates one of the aspects of *natura naturata*. This is the Earth within its objective limits, objectified by *Homo faber*, the instantaneous Earth one defines by depriving it of the forces that inhabit it—biological, ecological, and geological forces that work endlessly to transform it from the inside. The Earth-object is the Earth, humanized by way of a suffusion of the human subject.

2. The neo-organicist Earth body is the animal that possesses its own power of creation and obeys laws that may not be the same as human laws. In other words, it is the Earth that expresses nature as *naturing (natura naturans)*, the Earth-subject. In contrast to the Earth-object, this attribute of the Earth reveals the powers of transformation that lead it to always being something other than itself: The *full body* of the Earth is ripe with what it will—potentially—become.

3. But there is also the Earth-withdrawn, nonobjective and irreducible to a body, the Earth whose being eternally eludes its spherical aspect. This is the Earth governed by denaturing nature: the nocturnal side of nature. Whenever

a photo of the Earth is taken, we always forget the recessed part that the image is not able to capture, we repress the fact that the Earth, as with any being, contains a part that is not only "unthinged" (Schelling) but also decidedly unthingifiable.[15] As Iain Hamilton Grant writes, through his reading of Schelling, the unthinged is the part of the thing that is always in becoming, always exceeding that which could contain it within the limits of an object—in other words, that which is endlessly emerging, the productivity of nature that always exceeds the limits *of* the product within the *product*, the naturing subject within and beyond the natured object.[16] We will therefore categorize as unthingifiable not the natural productivity that is expressed within the Earth but the antiproductivity that is expressed neither in the form of a naturing subject nor in the form of a natured object.

An Earth that is neither subject nor object: Is this not precisely what Hannah Arendt attempts to proclaim in her book *The Human Condition*? Initially published in 1958, Hannah Arendt maintains that that the Earth is the "very quintessence of the human condition"—not human nature (who or what the human being is) but that without which the human would merely be an abstraction, an *essence without existence*:

> An earthly nature, for all we know, may be unique in the universe in providing human beings with a habitat in which they can move and breathe without effort and without artifice.[17]

Her claim that the habitat conditions human beings means that everything they think and do would be completely different if they somehow lived on another planet. Of course, they would still be human beings, but their *way* of being would be different, if only because their living conditions would be first and foremost artificial and not given.[18] To pursue this claim, *The Human Condition* begins by questioning the relation between nature/artifice: Launched in space in 1957, for a brief moment in time, the first Sputnik satellite "managed to stay in the skies; it dwelt and moved in the proximity of the heavenly bodies as though it had been admitted tentatively into their sublime company."[19] The attempt will be considered a failure in the sense that *Sputnik* will only remain among the heavenly bodies for but a very brief moment; however, Arendt sees something much more interesting in this attempt: a "rebellion against human existence as it has been given," this "free gift from nowhere (secularly speaking)" that the human being "wishes to exchange, as it were, for something he has made himself."[20] For Arendt, the Earth is therefore not an object, like Sputnik is: We can't produce it. But the Earth isn't Gaia either, an instance of productivity or a supersubject: The Earth is the condition of possibility for the existence of human subjects, a *concrete transcendental* in the sense where it is simultaneously "quintessence" and "habitat." If the Earth is, on the dark side, unthingifiable, it's in the sense that if it has been produced, the result of human labor, crafted by *Homo faber*, having become an object, it no longer would be the Earth understood as the human condition, as *concrete transcendental habitat* given to man. At best, it would be the inhabited, the reduced-to-the-inhabited, even the machine for living in, to use Le Corbusier's term—*une machine à habiter.*

Here Arendt rediscovers Heidegger's analyses that we cited in the first part of this present work: The Earth viewed from outer space is no longer the "earth on which man lives," it is no longer the homeland (Heimat). In the same way that for Husserl, the Earth is not a heavenly body in motion among many others but first and foremost a "ground," it "does not move and does not rest," since "only in relation to it are movement and rest given as having their sense of movement and rest."[21] It is in this sense that the Earth is the *"arche-dwelling,"* the "ark of the world" that nothing can replace and that we reference and intellectually give as an answer each time we imagine that the Moon or an airplane could constitute another foundational basis or "ground."[22] "Arche-dwelling," "homeland," "habitat," for Arendt, Heidegger, and Husserl, the thesis is clear: humanity is *under condition* of the Earth, understood as that which can't be reduced to an object, or a subject—in other words, a transcendental nonobjective form. The problem is that the condition seems to continuously make a U-turn and fall back toward the conditioned: The Earth is only withdrawn in order to leave a place (an area for) humanity. The Earth is indeed not an object or a subject but prepares the world *for* the human subject and its objects. It almost seems to inherently bear what is necessary for humanity: It is the Earth on the *condition* that it is *for* human beings—as if the human condition became the condition according to which the Earth can exist, as if humanity provided itself with its own transcendental. The nocturnal side of the Earth—the human side—proves to be nothing but the eclipse of a humanity broadly projecting its shadow onto the surface of the globe. Human, all too human Earth.

THE EARTH AS UNCONSTRUCTABLE TRAJECT

Must we then rid the Earth of any consideration of the human? Are we striving to celebrate the inhuman, the coldness of a planet that has no concern for us?[23] Our theoretical ambition is somewhat different: We wish to show that the "arche-dwelling" (Husserl), "homeland" (Heidegger), and "habitat" (Arendt) circumscribe what would be judiciously called a *restrained anthropological* point of view, a point of view suffering from modern and hypermodern acosmism. We would like to go beyond this point of view and get rid of this acosmic anthropocentrism without denying that the relation human beings entertain with the Earth as a dwelling or place of residence, from *now on*, fully plays a part in what we mean by "the Earth."

In a certain way, Arendt's concrete transcendental is neither concrete nor historical enough, just as Husserl's arche-dwelling does not take into account the fact that the Earth has *not* always been *for* humanity. And Heidegger failed to make a cosmic event out of the "homeland." In order to concretize the transcendental of the Earth, we must not consider it as an object (that we can capture from outer space thanks to a camera) or as a quasi-subject (such as Gaia, a rather local expression of naturing nature) but rather as a trans-ject or perhaps more specifically and simply a *traject*, as an interval spanning space-time. In fact, the Earth is not merely a "ground" upon which we stand, not simply a planet surrounded by a moon and artificial satellites; it's also a long-term event that began 4.54 billion years ago, the historical trajectory of an entity that will disappear in several billion years. As an *eventual traject*, the Earth exceeds the Gaia hypothesis to the extent that it began first as a fiery star without life or as an atmosphere without an ecosphere. This geological event that lasts billions of years dwarfs the geo-constructivist pretension of remaking nature—unless you place the Big Bang or the formation of the solar system on the same plane as the climate geoengineering project dreamed up by NASA . . . As stellar historical trajectory, the Earth is a singular event that we would not be able to reproduce in a laboratory. As unconstructable traject, the Earth can clearly be neither controlled nor dominated. From its dark origins to its glacial ends, the Earth will always love to hide.

To speak of the Earth as an unconstructable historical trajectory is a way of escaping purely objective-spatial conceptions of the planet (as a box, a marble, ground, ark, etc.) so as to truly take into account a temporal dimension.[24] This expansion of the temporal dimension is, of course, one of the effects of the Anthropocene on the contemporary theoretical field, but what we're attempting to do here is grasp a relation with the Earth that goes "beyond human experience."[25] We can nevertheless ask ourselves how this extreme temporal opening can indeed inform reflections on ecology: Is it not more important to be concerned with life on Earth instead of musing about these ancient times when life didn't yet exist or those times in the distant future when life will no longer exist? A number of contemporary thinkers have conceived of a world "without us" (Eugene Thacker, Alan Weisman),[26] an "earth after us" (Jan Zalasiewicz),[27] an ancestral "Earth" revealing a "world where humanity is absent" (Quentin Meillassoux).[28] However fruitful these reflections about a world without thinking subjects are, they still present us with two problems:

1. These theorizations seem to be, whether consciously or not, the inverse of the ecological reality upon which they rely. Above all, the anaturalist drive

actually attempts to create an *us without a world*, a humanity dreaming of the possibility of extracting itself from an already ravaged ecosphere. In other words, these theorizations skip the step that would allow them to understand the ecological situation produced not by an absence of humanity but by its omnipresence. The theorizations regarding a world without us don't help us to think the acosmism resulting from humanity's omnipresence.

2. Admittedly, they eliminate the anthropocentrism that we have described as too restrained (too centered on a specific moment of the Earth's history: the history of humans), but in being caught up by a decentering venture that is ultimately destructive. While they confirm what Lewis Mumford calls "the crime of Galileo,"[29] they completely ignore what remains for us to establish: not the abrupt annihilation of any place for humans but an *anthropology commensurate with the universe*—that is to say, opened up by the universe and that sets in motion a modification of our relation with the Earth instead of the abrupt annihilation of any place left to humans. It is therefore insufficient, and clearly nihilistic, to desperately attempt to promote a theory of the Earth that is based on an Earth prior to life and animal consciousness, the Earth as the object of a mathematically formulated knowledge without any subjectivity whatsoever. This fiercely unilateral conception maintains the divide between "us" and the "world," between the human "subject" and the "ancient" realities that can only be accessed by pure science.[30] It would be more accurate to show that the Earth, as a long-term traject, as an event arising from a far-off place, integrated human life and humanity once they appeared.[31] The Earth did not produce the living. The Earth is not a quasi-divine subject or Gaia. Nor is the Earth a product, an object, or a spaceship that one would be able to pilot. The Earth is an alliance of all the elements that compose it. From now on (and not since the very beginning), the Earth as traject (trans-ject) *traverses* the living and the nonliving, the nonconscious mineral and the diverse levels of the consciousness of the living; but the Earth does not *oppose* them. Henceforth—and until the world vanishes—the Earth bears and carries within itself the multiplicity of forms of human and nonhuman existence that have appeared not *on* Earth but *with* the Earth.

THE ECCENTRICITY OF THE EARTH AND THE POINT OF VIEW FROM THE OUTSIDE

Driven by the anaturalist tendency formatted by modernity, the hegemonic discourse of the Anthropocene is founded on the belief according to which it is possible to replace nature with an artifact—to replace our planet with a *Sputnik*, as Hannah Arendt melancholically notes in *The Human Condition*. From this point forward, the objective is to use geoengineering to replace the Earth with itself, an Earth with an optimized climate, with an augmented reality. It's not a question of colonizing another planet but—and this is perhaps a tautology with frightening consequences—of *terraforming the Earth*. The vitalist and organicist metaphors—that turn the Earth into a great big animal where each part is connected to the whole—serve to fuel philosophical and political oppositions to the project of geoclimatism and any form of geo-capitalism. These figures of speech allow us to restore the qualities of the nonhuman universe that anaturalism tends to erase (we spoke about this restoration in regard to Darwin). Furthermore, it seems completely logical to us to appeal to the term *nature* when it is used as an expansive synecdoche—i.e., when the whole stands for the parts. This back-and-forth between the parts and the whole, between the whole and its parts, indeed assures the vitality of terrestrial nature against its contemporary reductions. But the Earth, terrestrial nature, is not limited to this organicism and vitalism. As an irreplaceable event that cannot be reproduced in a laboratory, the Earth is a wild entity that, over billions of years, has dodged geo-constructivism's absurd lassos. Retreating into the most distant past and the most inaccessible future, hiding under the open sky in its spatiotemporal trajectory, the Earth is the metonymy of denaturing nature. Consequently, the Earth is the concrete transcendental that, without reducing itself to something for humanity, accompanies humanity throughout the time that it will be granted.

The Earth reveals its utmost singularity in its relations with the living, and specifically with the living humans and their capacity for a vast anthropology, commensurate with an incommensurate universe. The Earth is most exceptional in its joining the living with the nonliving and the Anthropocene present with the still-smoldering memory of the Big Bang that remains in the Earth's molten core. As the paleobiologist Jan Zalasiewicz notes, the Earth's singularity is derived more from its quality of being a "muddy planet" than a living planet, a mud composed of the chemical alteration of soils, of the decomposition of the living—in other words, composed of the permanent reassemblage of the Earth subject to the *encounter* of the living with the nonliving, of their history and the endless stratification of this history.[32] In 1975, the philosopher Hans Blumenberg also addressed this singularity:

> A decade of intensive attention to astronautics has produced a surprise that is, in an insidious way, Pre-Copernican. The Earth has turned out to be a cosmic exception.[33]

If the initial outcome of the Copernican revolution was "centrifugal," its real and substantive effect ultimately will have been "centripetal." Within the infinite Galilean universe, the Earth appears more and more as a *cosmic eccentricity*. That the Earth is eccentric means:

1. The Earth in no way forms the center of the universe and is therefore outside of any central position.

2. The Earth is eccentric in the same sense of when we speak about someone who is original, who does not act like everyone else. So, the Earth is not the center of the universe? Then it is all the more singular and irreplaceable.[34]

And yet, we became aware of this eccentricity of the Earth thanks to technologies humans launched into space. It would be disastrous to refuse all forms of technologies in the name of some sort of neo-organicism, but it would be just as dangerous to proceed, as Levinas, in praising the technology that "wrenches us out of the Heideggerian world and the superstitions surrounding *Place*": "For one hour," Levinas writes in regard to Yuri Gagarin,

> a man existed beyond any horizon—everything around him was sky, or more exactly, everything was geometrical space. A man existed in the absolute of homogeneous space.[35]

But no living being can "exist" in a "geometrical space," in such a space no "human face" can "shine in all its nudity." If for Levinas, the "enemies of industrial society are in most cases reactionary,"[36] it's because he didn't want to truly comprehend that it is *the Earth—a "Place"—that has been viewed from the outside*: Pace Levinas and following Blumenberg, we argue that Gagarin's experience does not confirm but rather contests Galileo's homogeneous space. Accordingly, we must question at length our relation to the various *technologies of eccentricity*. Will astronautics have been used to undo terrestrial nature, as Arendt and Heidegger suggest in their critical stances (and, paradoxically, Levinas as well with his positive description of Gagarin's geometrical space)? Or can astronautics, as a post-Galilean science, help us to contemplate our terrestrial nature in different terms? For Blumenberg, it all depends on the type of experience human societies decide to symbolize:

> It's only as an experience of turning back that we shall accept that for man there are no alternatives to the Earth, just as for reason, there are no alternatives to human reason.[37]

On the face of it, geo-constructivism is based on the opposite experience: not returning to Earth. Remaining in outer space, dreaming of the becoming-astronaut of the *anthropos*. Becoming without ever coming back is one of the aspects of the contemporary fantasy that no accelerationist, no transhumanist, and ultimately no constructivist will venture to call into question. However, instead of a generalized becoming-astronaut—or specifically a becoming-astronaut reserved for those who can afford to purchase their ticket—we could be the first beings capable of a perspective that comprises the point of view of the outside. If, as Anselm Franke claims, the "Earthrise" photos of the 1960s and 1970s led to a disappearance of the extraplanetary outside (as cameras reversed their focus to turn inward, back toward Earth), it now seems possible to reinstate the place of the outside—on behalf of the Earth's singularity, which is now accentuated and dramatized by the precarity of human habitats.[38] *Sputnik* and the other Apollo missions must be considered not as a means for leaving Earth in search of a replacement planet but as the best technologies for recognizing the Earth's radical eccentricity.

VISIONS OF RETURN (THE DIALECTICS OF *GRAVITY* AND *INTERSTELLAR*)

It would seem that we have uncovered our ecologist and environmentalist password and that the conclusion of this present work can now be announced with great fanfare: Return to Earth! We must not, however, be too eager, as we don't want the return to result in a disastrous crash upon re-entry into the Earth's atmosphere. In fact, this return is anything but a new idea; it is the comet's tail of a very bold earlier project. These new ecology-explorers are likely to return to Earth only to find that the geo-constructivists are already there, or to be more precise, the geo-constructivists had *overtaken*—outdistanced and remade— them. We should not forget what we've already made clear: The *reversal* of the frontier consisted precisely of an imaginary and financial repatriation of extraplanetary space toward the Earth. With the Anthropocene, geo-technological capitalism has already invaded the whole territory. That's the inherent ambiguity of Mumford's epigraph for the second volume of his *Myth of the Machine*:

> If the key to the past few centuries has been "Mechanization takes control," the theme of the present book may be summed up in Colonel John Glenn's words on returning from orbit to Earth: "Let Man Take Over."[39]

For Mumford, this statement heralded a moment for a healthy critique of mechanization; but today, we know that humanity's initiative to take back control has only led to an unprecedented reinforcement of mechanization—an abstract mechanization that has become the virtual essence of the world, of a rearranged nature. However, we could easily sidestep this argument: The return of the geo-constructivists is a return to the Earth object, that maintains a form of extraterritoriality. From outer space, the geo-constructivists expect nothing more of the Earth than a *return on investment*. From their point of view, what is essential is not trying to reconsider a new way of inhabiting the Earth but rather a new Earth to inhabit—once the Earth has finally been augmented. Nevertheless, how can we prevent inhabiting the Earth object? Latour proposes the following solution: We must "come back to Earth";[40] but this return is on the condition of humankind's capacity for accepting to be "taken by" Gaia, of accepting the *"coming back down to Earth"*[41] as an "intrusion" of Gaia, who, terrifyingly, "is coming towards us."[42] And yet, what exactly is Gaia for Latour? Certainly not the Earth, since "With its finger, quite simply, Gaia designates the Earth."[43] Gaia is another name not for the Earth but for horror. And all Latour clearly sees is a way for calming this anguish: "Bow before Gaia's majesty."[44] Consequently, how should we interpret this feeling of horror that is present from the very first pages of *Facing Gaia* and that traverses the entirety of the book? Gaia is the return of the repressed. Indeed, contrary to Dr. Frankenstein, Latour has developed the idea according to which we should love our technologies and not flee from the horror in confronting them. But here we have Dr. Latour who is going to see all the consequences of his thought surge back over him: Gaia, product of an unconditional love, is a figure of terror that inspires no love whatsoever in Latour. And yet, Gaia is the product of the economico-technological reformation of the Earth. The entire problem is revealed to us by way of the term *intrusion*, which Latour uses following Stengers. That Gaia is hurtling toward us and proves to be intrusive is a consequence that masks—by way of a good eco-pragmatist logic—the initial cause: the technological extrusion of humanity into the sky. The intrusion of Gaia is the backlash.

The use of the term *intrusion* should be read as the symptom of a conceptuality closed in on itself, since eco-constructivism is characterized by a refusal of all exteriority (since everything is attached, tied together, hyperconnected, etc.). If we share Latour's critique of the transcendent viewpoint, of "the point of view from nowhere"[45] that we can attribute to geo-constructivist astronautics, we have also insisted on the danger of the inverse viewpoint translated as the foreclosure of the outside. This foreclosure transforms *strangers into danger*— "foreigners can put you in a situation such that your existence will be denied," Latour writes.[46] As if every true foreigner, and not merely the one who takes up language diplomatically in a "Parliament of things" (*Politics of Nature*) or a "theater of negotiations" (*Facing Gaia*), ran the risk of depressurizing the eco-constructivist world without an outside. In other words, we don't want the return to Earth to result in a funeral. In order to be sure that's not the case, we must be able to metabolize this nowhere, of accepting the nomadism of those who do not reside in definitive territories: The return to Earth must coincide with a return of the outside.

We can help clarify this latter formula by way of a dialectical interpretation of two recent films, *Gravity* (dir. Alfonso Cuarón, 2013) and *Interstellar* (dir.

Christopher Nolan, 2014).[47] At first glance, it may seem that *Gravity* should be deemed the ultimate environmental and ecological film par excellence: a woman returns to Earth and is physically and psychically reborn, after having lived in the zero gravity of artificial satellites in space. In stark contrast, the film *Interstellar* clings to a sort of environmentalist ideology responsible for the refusal of developing technologies; this ideology goes so far as to hide the reality of the lunar landings from newer generations so as to force humanity to only care about the Earth. Thanks to a clandestine NASA, humans will finally end up launching out into space in order to find a suitable replacement planet—according to the logic of "Plan C" evoked in the first part of the present book. *Interstellar* advocates the theory of the "frontier" through the very mouth of the main character from the film:

> It's like we've forgotten who we are: explorers, pioneers, not caretakers. We used to look up in the sky and wonder about our place in the stars. Now we just look down and worry about our place in the dirt.

It should come as no surprise that Nick Land, an accelerationist opposed as much to any leftist position as to any sort of environmentalist discourse, lauded the merits of the film;[48] but we will not, however, conclude that it would be fruitful to simply reject this *point of view of the outside*. It seems necessary for us to dialectize the two films: instead of rejecting what there is *between the stars* (the interstellar), we need to incorporate it along with the Earth as its *inner interval*, its alterity from outer space, so as to prevent the Earth from once again becoming a woefully compact and one-dimensionally flat star. If, at the end of *Gravity*, Ryan Stone (Sandra Bullock) is able to return to Earth (like a stone), and become reborn without repeating the same melancholic lifestyle that she led among the artificial satellites, if she no longer looks to shape the Earth like a Sputnik (Arendt's nightmare), it's because she learned how to adopt the *point of view from out there*.[49] And she learned how to conserve this strangeness on Earth, its definitive atopia.[50] She will know that the territory we inhabit is never simply a territory but always *more* than *this* territory—never simply the defense of its particular interests, never simply in competition or infringement with another territory. Each territory must be the bearer of this unconstructable Earth that spins in the Night, this Earth that requires less a personification (Gaia) than metaphors—i.e., in the etymological sense, transports, passages, above the ruins of time. We are not advocating for a return bound to a territory closed in on itself, severed from everything that could be aterritorial, but the return of an aterritoriality establishing the Earth as a traject, the dismantled wild foundation that boundlessly differentiates the territories of the world.[51]

CONCLUSION

What Is to Be Unmade?

It is the immediate *task of philosophy*, which is in the service of history, to unmask self-estrangement in its unholy forms. . . . Thus the criticism of heaven turns into the criticism of earth.
—Marx, *Critique of Hegel's Philosophy of Right*

Today space is splendid
The mountains have come loose
Let's unmake something
—Michael Palmer, "Odd-Even"

THE EXTRA-GEOLOGICAL DIMENSION OF POLITICAL ECOLOGY

From the Copenhagen Summit in 2009 to the vision of a transitory Earth, carrying along with it the aftereffects of the Big Bang, and destined to be engulfed by the red giant that the Sun will become in barely five billion years, it would appear that we have led ourselves astray from the more immediate environmental and ecological concerns of our world and that any idea of political ecology scrupulously studied for the duration of the universe seems vain. What does it mean to preserve an Earth that is destined for cosmic destruction? No doubt, nothing. If not simply to fully hand it over to the amateurs of "plan C," to all those who dream of colonizing the outer edges of the solar system.[1] Nevertheless, our inquiry has simply prolonged that which, from now on, the notion of the Anthropocene forces us to do: work on very long-term durations. Even the sudden disappearance of humanity will not put an end to the Anthropocene—that is, put an end to the consequences of current anthropogenic activities at the scale of several hundreds of thousands of years. In this sense, our extreme sensitivity to worlds-without-us is nothing more than an echo of the global ecological condition and the eventual premature disappearance of humanity that will have nothing whatsoever to do with the extinction of the Sun. If it is useful for us to consider worlds-without-us, it shouldn't lead us to obscure or mask the way in which we produce the us-without-worlds.

Consequently, we must invent a political ecology that simultaneously has a firm grip on the world and that is also capable of granting space for an extra-geological dimension that the analyses of the Earth as traject[2] have provided us. Should we claim that what is at stake is forging a political ecology commensurate with the universe? This formula echoes that of Bataille, who attempts to establish "an economy commensurate with the universe" that would not have at its core the conservation, acquisition, and growth of beings and the search for utility but expenditure, loss, useless consummation and consumption, all that extravagantly provides in the manner of the sun.[3] Certainly, geo-capitalism itself takes up the task of the dilapidation of resources, but—in complete contrast to what Bataille was looking for—in no way does this dilapidation lead to some kind of ecstasy, excess, or radiance: On the contrary, it produces a collapse. To avoid this collapse, we need a political ecology capable of grasping the geological problem by its roots in placing the universe at the core of the ecology. From now on, political ecology must deal with the obscure foundation of nature, the black sun of antiproduction, the unconstructable base that makes up the trajectory [*trajet*] of the Earth. But who are the political ecologists that can help us to concretely formulize this new political ecology? How can a critique of the anthropocenic sky lead us to a political critique of the Earth?

GEOLOGICAL CAPITALISM AND ITS ENEMIES

The analysis of geo-capitalist logic is absent from Latour's analyses, who carefully avoids the name of Marx—as well as the names of André Gorz and Ivan Illich. And yet, the lessons provided by these latter two thinkers are still relevant: A political ecology that doesn't begin by assuring itself of an adequate critique of capitalism can easily justify the place for a "managerial fascism"[4] (Illich), of a military management of rations that will maintain as long as possible the abyss between the wealthy and the disadvantaged.

Concerning this point, André Gorz makes the following claim: A critique of capitalism inevitably leads to accounting for the environmental disasters it provokes. However, starting from the principal consideration of the ecological imperative, of the lone defense of the "living environment," or—as one would say today—of Gaia, can lead just as much to anticapitalist options such as a "green Pétainism, an eco-fascism, or a naturalistic communitarism."[5] But what is the form of contemporary capitalism? Expertise, technocratism, *top-down management*, geoengineering, Earth stewardship, a stewardship founded on an ecology of resilience, the demand for adapting to the ontological chaos that is so often nothing more than the cover for the chaos produced by the economic system: Geopower is the promotion and management of geo-capitalism, if this is the name we give to a form of capitalism propped up by the reformation project of the Earth and the synthetization of life. Unearthing fossil fuels resulting from the degradation of organic material hidden within the soil for millions of years, harnessing living beings from microbes to human beings in passing by the animals of the meat industry, not only does geo-capitalism exploit the living whether still alive or dead, but it attempts to recreate and produce a synthetic environment: The preservation of the "living environment" is less and less the option chosen by geo-capitalism and its option of post-preservation. If, as Gorz writes, the capitalist system "deprives individuals the possibility of having—and sharing—a world," if this system expropriates and marginalizes beings from their "lifeworld,"[6] what needs to be posed again in a new and fresh way is the question of the political reconquering of autonomy, to the extent that the ecofascism of tomorrow will not strive so much to maintain the purity of beings and territories as to modify and increase it just in time.

As such, it seems problematic for us to consider, along with Latour, that the enemies will be "humans," defined by way of their ungrounded position, and that we should remain on the side of the "Earthbound," ready to "bow" in facing Gaia.[7] Firstly, because this war is already led by promoters of extreme energies, promoters who don't hesitate lashing out at any humans who obstruct their access to fossil fuels hidden deep in the Earth. It is also led by theorists of transhumanism and their sinister postnaturalism. Our aim is not to heap more hatred onto humans. And secondly, because Latour's Earthbound can be identified through their capacity for expressing an interest and for delegating legitimate spokespeople for speaking in the interest of the soil, the atmosphere, the cities, etc.[8] This logic of territorial belonging that defines the Earthbound does not mention that the territories themselves are always divided and never compact. One can thus defend a territory, in a violent reactionary way, from migrants or radical environmentalists as has recently been the case in France by certain farmers and members of the extreme right. Following a perfectly neoliberal reasoning, we can also wish for "setting territories into motion" and "explore the possible futures of the national territory in order to conceive a politics of planning and zoning adapted to the identified future challenges and stakes."[9] The enemies are, in each specific situation, what allows us to understand that a local issue always refers back to a determined politics and economy. As such, when in May 2015 civilians of Burkina Faso marched to

Ouagadougou in protest against the GMOs produced by Monsanto,[10] it wasn't primarily the defense of the territory that was at stake but a fight against extractivism engaging in the possibility of living in a specific territory.[11] One must choose between kneeling or confronting.

POLITICAL ECOLOGY OF SEPARATION: THE WORK OF UNMAKING

Against the model of endless growth realized through the extraction of extreme energy sources, we must be able to propose a societal model that would cease simply targeting development for the beneficiaries of the Anthropocene and to the detriment of those who, in the Global South as well as certain parts of the Global North, suffer its consequences. Nothing speaks more to this attempt than the mechanism titled REDD (Reducing Emissions from Deforestation and Forest Degradation), a mechanism of the carbon market that allows industries and wealthy nations to compensate for their environmental degradations through the purchase and pseudo-protection of forests in Africa or Latin America. Such a mechanism has as one of its direct consequences the displacement of populations—for example, the Sengwer people of Kenya—in order to preserve "protected" vacant spaces. It's very clear, it's not at all a question of preserving natural environmental spaces, but, as Nnimmo Bassey explains, about allowing for the polluters to continue to pollute, to "secure" carbon: "In the forests, the trees are seen as carbon stocks, and not as trees anymore."[12] On one hand, untouchable stocks that force farmers to abandon subsistence farming in order produce carbon-stock trees. On the other hand, a soulless destruction. We can understand why the famous 2010 People's Agreement of Cochabamba rejected without any ambiguity such a financial mechanism:

> We condemn market mechanisms such as REDD (Reducing Emissions from Deforestation and Forest Degradation) and its versions 1.0 and 2.0, which are violating the sovereignty of peoples and their right to prior free and informed consent as well as the sovereignty of national States, the customs of peoples, and the rights of nature.[13]

Counter to this model of split growth that destroys on the one hand while accumulating on the other, it is necessary to propose a model of de-growth. Gorz, who began to use this term in the middle of the 1970s,[14] maintains that the end of growth would be inevitable: either imposed from above, according to a heteronomic model, or invented in an autonomous way—i.e., as something desired. On one hand, a logic of "sacrifices" that will, and we have no doubt about this, be implemented against the poor and, on the other hand, we have the eventuality of an "enforced renunciation."[15] To renounce means to abandon (the use, the enjoyment of something), to undergo abnegation, but also to *announce a return (re-nuntiare)*, thanks to a kind of *backward movement* suggested by the prefix *(re-)*. In this sense, renunciation is a form of *resilience through separation*: to undo, to withdraw, but in order to, in the end, announce or propose another vision of the world. Renunciation in search of a new meta-phorization of the world implies firstly this necessary *distancing* that an ecology of separation produces, that doesn't consider geo-capitalism as a fatality. This detachment allows one to place the model of hypermodernity under contingency, to renounce the subjects that we were and their relation to the world, with one's sights set on announcing in return another configuration of existence. De-growth can therefore only mean one and one thing alone: a radical change of the community of the world-society.[16]

Our thesis maintains that such a change implies an ontological reorientation. By this expression, we mean contesting not only the productivist model but the category itself of production. Throughout the course of the present work, we have criticized the constructivism at work in the geo-constructivist and eco-constructivist discourses and practices—these two forms of constructivism are not identical, and nothing prevents us from envisioning the existence of other forms of constructivism.[17] Eco-constructivism is this optimistic ontology that believes in the capacity of supra-atomic entities—objects, animals, humans—of assembling themselves under the modality of uncertainty. Its ideal is the immediate magic of a world without an outside, where everything is full of agency. Geo-constructivism exploits eco-constructivist thought, it appropriates eco-constructivist thought for itself in order to justify its project of reformatting the Earth: It occupies and saturates the outside (i.e., the blind spot of eco-constructivism) and manages—imaginarily *off-planet*—the hell of fusional hybridizations and the *slow dilapidation* of the world that the dominant eco-constructivism accompanies. By *slow dilapidation*, we mean the process generated by the conception of products that are more and more complex: made from alloys and composites, from elements that are more and more miniaturized (ending up at the level of nanotechnologies), these products become impossible to recycle since, as the engineer Philippe Bihouix writes, we cannot "locate the different metals and separate them out."[18] Unrestrained hybridization leads to so-called "dispersive" or "dissipative" uses of the metals used as chemical additives in products (in plastics, inks, cosmetics, etc.). In this sense, an ecology of separation leads to contesting the mode of production that results in the slow dilapidation: It requires a restraint in production—an antiproduction contesting the positive magic of eco-constructivism and the negative technology of geo-constructivism.

But, in concrete terms, what does antiproduction mean? How can we help denaturing nature enter into the equation of political ecology? The least of things we can do is to try to accompany the energetic "transition" of cities—we should think here of the transition town movement that aims to reconfigure cities and towns according to a "postcarbon" economy.[19] We also believe in the necessity of forgoing disposables, and Bihouix is correct by insisting on the necessity of a *return of refundable deposit* (on reusable bottles, for instance) that would force us to take care of what we use and remind us that what is produced must *return* to the state prior to its being produced—what we have called the nonthingifiable. And yet, the ontology of production makes it very difficult to think that a return is not merely a simple way to return and stand on a territory, but it defines the *otherside of becoming* and accompanies each entity all the way until its end. The foreclosure of the endless movement of return and the fetishism of becoming makes what Illich calls the *"recourse to precedent"*[20] impossible (i.e., the recourse to that which exists before and, from this point forward, beyond the product of the present instant).

Let's express this idea in a more general manner: The project governing geo-constructivism and geo-capitalism is a project of *synthesis*—of assemblage, composition, etc. The science that instructs geo-power is not so much looking to discover but to *conceive of all the pieces*, which is what it proposes under the guise of *pieces of evidence*. By contrast, an ecology of separation founded on the imperative of antiproduction is an analytic ecology that thinks the possibility of separation (of recycling) instead of precipitating itself toward a fusion. Where geo-constructivism tasks itself with remaking the Earth, we must learn to detach ourselves from the synthetic fantasy. Detach ourselves from a certain mode of civilization that either smothers us within networks of immanence or strives to expel us far from any ground whatsoever into outer space where we dream of a frozen body that will never melt. Detach ourselves from the idea according to which a miraculous "reflexivity," thanks to the help of more technology, will allow us to resolve the problems arising out of a technophilia—or the opposite, that the immersion into Gaia or her veneration will save us from our sin of being human. Not remaking—unmaking. Far from simply being capable of being reduced to an abstract word, unmaking could define a political verb for the minoritarian bodies of the Anthropocene. In terms of environmental justice, unmaking concretely means to dismantle nuclear power plants once they have become dilapidated, or more profoundly, it means to abandon nuclear power as such, a project that is not only ecologically dangerous but just as antidemocratic. But the energy question cannot be reduced to nuclear power: At the time I'm writing this, the Standing Rock Sioux Tribe of North Dakota is rising up and confronting the construction of an oil pipeline that will run through part of their territory on Lake Oahe along the Missouri River.

Consequently, to unmake always defines two operations. One is intellectual and invites us to rid ourselves of any of our past illusions, such as those related to Progress or to "cheap nature" (Jason W. Moore). The other is practical: To unmake means to dismantle or prevent from constructing that which harms us. These two operations are effectuated in reverse of geo-capitalist production, which produces as if antiproduction didn't exist, extracting fossil fuels from the depths of the Earth as if nothing could restrain them. This extraction has not stopped since the passing of the Paris Climate Accord in December 2015: To contain the warming of the planet at under 2 degrees Celsius (or more specifically under 1.5 degrees, which is now required) we need to forbid all new extraction projects—as such, we must reconsider and reflect upon an entirely new future.[21] Today, voices can be heard demanding that fossil fuels remain in the ground. Spoken by the minoritarian bodies who have suffered through climate changes more than anyone else, these voices perfectly express the planetary and extraplanetary logic—commensurate with the Universe—that political ecology demands today: Leave fossil fuels in the Earth and refuse the practice of geo-capitalist synthesis. This refusal consists of maintaining under our feet the solar energy that has been captured by way of photosynthesis and that has declined with the decomposition of plants. Through this material, economic, and political separation, a symbolic relation is established with what is dead. This energetic potential that will remain in the Earth will not be a sign of death; it is at every moment the active memory of what is at the origin of terrestrial life: the Sun, as metonymy of the universe into which the Earth has been cast. Life and death under our feet, this is how a planetary ecology concerned with the future of humans and nonhumans can leave a place for the inhuman.

ACKNOWLEDGMENTS

I would like to thank Christophe Bonneuil for his advice and for the confidence he had in helping to make this book a reality. I would also like to thank Sophie Gosselin for her crucial comments as well as François Roussel and Sophie Gouverneur. And finally, I want to thank Monique Allewaert, who accompanied me through every stage of its completion.

NOTES

INTRODUCTION: RECONSTRUCTING THE EARTH?

1. Michel Tibon-Cornillot, "La reconstruction générale du monde," *La Planète laboratoire* 1 (2007): 4.

2. Mark Lynas, *The God Species: Saving the Planet in the Age of Humans* (Washington, DC: National Geographical Society, 2011), 8.

3. Ulrich Beck, *La Société du risque* (Paris: Champs-Flammarion, 2001), 22, 340–98.

4. Beck, *La Société du risque*, 23.

5. Günther Anders, *L'Obsolescence de l'homme*, vol. 2, *Sur la destruction de la vie à l'époque de la troisième révolution industrielle* (Paris: Fario, 2011), 397.

6. Hölderlin's equation in his poem "Patmos" is: "Where there is danger, / The rescue grows as well" (trans. Scott Horton, *Harper's Magazine*, July 16, 2007).

7. Erle Ellis, "Overpopulation Is Not the Problem," *New York Times*, September 13, 2013.

8. Philippe Descola, *Beyond Nature and Culture*, trans. Janet Lloyd (Chicago: University of Chicago Press, 2014).

9. Paul Feyerabend, *Philosophy of Nature* (Cambridge: Polity, 2016).

10. Aristotle, *Physics*, ed. David Bostock, trans. Robin Waterfield, Oxford World Classics (Oxford: Oxford University Press, 2008), 51. "And in general human skill either completes what nature is incapable of completing or imitates nature."

11. Here, we are supported by the findings in Patrick Blandin, *De la protection de la nature au pilotage de la biodiversité* (Versailles: Quae, 2009), 18–19, 28–30; Jacques Lepart and Pascal Marty, "Des réserves de nature aux territoires de la biodiversité. L'exemple de la France," *Annales de géographie* 115, no. 651 (2006): 487, 497, 503–4; Christophe Bonneuil, "Une nature liquide? Les discours de la biodiversité dans le nouvel esprit du capitalisme," in *Le Pouvoir de la biodiversité: néolibéralisation de la nature dans les pays émergents*, ed. Frédéric Thomas and Valérie Boisvert (Paris: IRD-Quae, 2015), 193–213.

12. Henri Atlan, "La Librairie du xxie siècle," in *L'Utérus artificiel* (Paris: Seuil, 2005).

13. Lewis Mumford, *The Myth of the Machine: Technics and Human Development* (New York: Harcourt, Brace & World, Inc., 1966), 3.

14. For more about the concepts of the biosphere and noosphere, see Wladimir Vernadsky, *La Biosphère* (Paris: Diderot Editeur, 1997), as well as Wladimir Vernadsky, "The Biosphere and the Noösphere," *American Scientist* 33, no. 1 (January 1945). Teilhard de Chardin is the one who invented the concept of the noosphere. See *La Place de l'homme dans la nature* (Paris: 10/18-UGE, 1956), 25, 109–72. See as well Frédéric Neyrat, *Biopolitique des catastrophes* (Paris: MF, 2008), 77–97.

15. Catherine Larrère and Raphaël Larrère, *Penser et agir avec la nature* (Paris: La Découverte, 2015).

16. See Jade Lindgaard, "Le grand bétonnage, une bombe climatique," *Mediapart*, July 27, 2015.

17. Eduardo Viveiros de Castro, "Et si le temps était venu de 'devenir indiens'?," *L'Obs*, July 12, 2014, http://bibliobs.nouvelobs.com/essais/20140710.OBS3375/le-temps -est-venu-pour-nous-tous-de-devenir-indiens.html.

18. Déborah Danowski and Eduardo Viveiros de Castro, "L'arrêt de monde," in *De l'univers clos au monde infini* (Paris: Dehors, 2014), 256. For more on this concept of a "humanity without world," see the book by the same authors, *The Ends of the World* (Cambridge, UK: Polity, 2017).

19. Paul Crutzen, "Geology of Mankind," *Nature* 415 (2002): 23.

20. Paul Crutzen, "Albedo Enhancement by Stratospheric Sulfur Injections: A Contribution to Resolve a Policy Dilemma? An Editorial Essay," *Climatic Change* 77 (2006): 211–20.

21. For more on this topic, see "Big Names behind US Push for Geoengineering," *The Guardian*, October 6, 2011; "Obama Takes Bold Step to Geoengineer Climate Change," *Huffington Post*, January 4, 2014. As for the *Wall Street Journal*, see Bjørn Lomborg, "Can Anything Serious Happen in Cancun? The Upcoming Climate Summit Promises More Proposals That Ignore Economic Reality," *Wall Street Journal*, November 11, 2010.

22. IPCC, "Summary for Policymakers," in *Climate Change 2013: The Physical Science Basis. Contribution of Working Group I to the Fifth Assessment Report of the Intergovernmental Panel on Climate Change* (Cambridge: Cambridge University Press, 2013), 29.

23. "Geoengineering—A Tool in the Fight to Tackle Climate Change, or a Dangerous Distraction?," *Huffington Post*, September 11, 2012. The article was written by Jon Taylor, then director of the Program for Climate Change at WWF-UK.

24. For more on the Heartland Institute and the American Enterprise Institute, which both reject climate change science but endorse *climate engineering*, see Clive Hamilton, *Earthmasters: The Dawn of the Age of Climate Engineering* (New Haven, Conn.: Yale University Press, 2013), 90.

25. "Scientists Urge Global 'Wake-up Call' to Deal with Climate Change," *The Guardian*, February 10, 2015. We learn that the CIA is also implicated in research on climate control. See "Spy Agencies Fund Climate Research in Hunt for Weather Weapon, Scientist Fears," *The Guardian*, February 15, 2015.

26. Suvi Huttunen, Emmi Skytén, and Mikael Hildén, "Emerging Policy Perspectives on Geoengineering: An International Comparison," *The Anthropocene Review* 2, no. 1 (2015): 14–32; John Virgoe, "International Governance of a Possible Geoengineering Intervention to Combat Climate Change," *Climatic Change* 95 (2009): 103–19.

27. Richard Buckminster Fuller and Jaime Snyder, *Operating Manual for Spaceship Earth* (Baden, Germany: Lars Müller Publishers, 2008), 56.

28. For more on the relationship between the imaginary of the Cold War and the "de-Earthed" [déterrestrée] vision of the planet Earth, see Christophe Bonneuil and Jean-Baptiste Fressoz, *The Shock of the Anthropocene*, trans. David Fernbach (New York: Verso, 2016), 58–64.

29. "Space Ark Will Save Man from a Dying Planet," *The Times*, April 28, 2014.

30. Matt Ridley, "Fossil Fuels Will Save the World (Really)," *The Wall Street Journal*, March 16, 2015.

31. Michael Shellenberger and Ted Nordhaus, *The Death of Environmentalism: Global Warming Politics in a Post-environmental World*, 2004, http://www.thebreakthrough.org /images/Death_of_Environmentalism.pdf.

32. See "An Ecomodernist Manifesto" (http://www.ecomodernism.org/), which we will examine in the second part of this present book.

33. For more on the terms *nonmodern* or *amodern*, see Bruno Latour, *We Have Never Been Modern*, trans. Catherine Porter (Cambridge, Mass.: Harvard University Press, 1993), 47.

34. Alex Williams and Nick Srnicek, "Manifeste accélérationniste," *Multitudes* 56 (2014): 34.

35. See Crawford S. Holling, "Resilience and Stability of Ecological Systems," *Annual Review of Ecology and Systematics* 4 (1973), and, more recently, the work of Lance H. Gunderson and Crawford S. Holling, "Resilience and Adaptive Cycles," in *Panarchy: Understanding Transformations in Human and Natural Systems*, ed. Lance H. Gunderson and Crawford S. Holling (Washington, DC: Island Press, 2002), 25–62.

36. Ilya Prigogine and Isabelle Stengers, *La Nouvelle Alliance. Métamorphose de la science* (1979; repr., Paris: Folio-Essais, 1986). See also the English translation of the same book, *Order Out of Chaos: Man's New Dialogue with Nature* (New York: Bantam Books, 1984).

37. Chapter 10 of *Process and Reality* that begins with "That 'all things flow'" and that examines "the subordination of fluency" in Descartes, Plato, and Aristotle. Alfred North Whitehead, *Process and Reality*, 2nd ed. (New York: Free Press, 1979).

38. Romain Felli and Noel Castree, "Neoliberalising Adaptation to Environmental Change: Foresight or Foreclosure?," *Environment and Planning A* 44 (2012): 1–4.

39. See Jeremy Walker and Melinda Cooper, "Genealogies of Resilience: From Systems Ecology to the Political Economy of Crisis Adaptation," *Security Dialogue* 42 (2011): 143–60; Melinda Cooper, "Turbulent Worlds: Financial Markets and Environmental Crisis," *Theory Culture Society* 27 (2010): 167–90; and Frédéric Neyrat, "Economy of Turbulence: How to Escape from the Global State of Emergency?," *Philosophy Today* (2015): 657–69.

40. Bruno Latour, *Facing Gaia: Eight Lectures on Climate Change*, trans. Catherine Porter (Cambridge, UK: Polity, 2017), 144.

41. "If a gap exists between moral principles and the world that we experience, it's our principles that are not moral and not the world." Émilie Hache, *Ce à quoi nous tenons* (Paris: La Découverte, 2011), 54–55. Eradicating all "gaps" between the two, in our opinion, would lead to generating another problem rather than a solution.

42. Bruno Latour, "Love Your Monsters," in *Love Your Monsters: Postenvironmentalism and the Anthropocene*, ed. Michael Shellenberger and Ted Nordhaus (Oakland, Calif.: Breakthrough Institute, 2011), Kindle.

43. For another way of interpreting this paradox see Claire Colebrook, "What Is the Anthropo-Political?," in *Twilight of the Anthropocene Idols*, by Tom Cohen, Claire Colebrook, and J. Hillis Miller (London: Open Humanities Press, 2016), 86–89.

44. Concerning the "monstrous fusions" of a "cyborg utopia"—a fusion of the organism and technology, of nature and artifice, of the given and the acquired, etc.—see Geneviève Azam, *Osons rester humain. Les impasses de la toute-puissance* (Paris: Les Liens qui libèrent, 2015), 89, 19. Nevertheless, we should add that these fusions are always *asymmetrical*, benefiting a colonizing term to the detriment of a subjugated term.

45. One of the principal thinkers of object-oriented ontologies is Graham Harman, who maintains that it is possible to define the essence of an object by way of subtracting it from all relational networks. *The Quadruple Object* (London: Zero Books, 2011).

46. William Cronon, "The Trouble with Wilderness," in *Uncommon Ground: Rethinking the Human Place in Nature*, ed. William Cronon (New York: W. W. Norton & Co., 1996), 69.

47. The so-called virginity of wilderness is from then on one of a trifold erasure: 1) a real erasure: the genocide of the native, out of which arises the impossibility of maintaining the cultivation of a nature proper to this tradition; 2) a technical erasure: it's often, rather late, at the moment of the creation of national parks that native people were actually displaced, as was the case with the Organ Pipe Cactus National Monument and the Organ Pipe Cactus Wilderness Area (see Mark Woods, "Wilderness," in *A Companion to Environmental Philosophy*, ed. Dale Jamieson [Malden, Mass.: Blackwell, 2001], 356–57); 3) a theoretical erasure: in other words, the desire to create a fiction of a nature outside of time and outside of history (see Cronon, "Trouble with Wilderness," 79–80).

48. One thinks, for example, of social Darwinism, applying the biological concepts coming from Darwin and evolutionary theory in regard to politics. See Patrick Tort, *Spencer et l'évolutionnisme philosophique*, Que sais-je? (Paris: PUF, 1996).

49. Thierry Jaccaud, "La vérité pour tous," *Thierry Jaccaud* (blog), January 10, 2013, http://www.thierry-jaccaud.com/2013/01/10/la-verite-pour-tous/.

50. Razmig Keucheyan, *La nature est un champ de bataille. Essai d'écologie politique* (Paris: Zones, 2014), 11, 17.

51. Keucheyan, *La nature est un champ de bataille*, 75–135.

52. Eduardo Viveiros de Castro, *Cannibal Metaphysics*, trans. Peter Skafish (Minneapolis: Univocal, 2014), 117, 74.

53. Viveiros de Castro, *Cannibal Metaphysics*, 66.

54. "Nature is the form of the Other as body" [La nature est la forme de l'Autre en tant que corps]. Eduardo Viveiros de Castro, "Perspectivisme et multinaturalisme en Amérique indigène," *Journal des anthropologues* 138–39 (2014): 171.

55. Félix Guattari, *Qu'est-ce que l'écosophie?* (Paris: Lignes/Imec, 2013). As we will show later on, Guattari's machinic ecology is supported by an ontological axiom: *Everything is production*.

56. Jamie Lorimer, who defends a "multinatural" approach of a biology of conservation, chooses the term *wildlife* as process and becoming at the detriment of the term *nature*. His defense of immanence, of nonhuman assemblages, of a "performative" preservation, of hybridity, etc., places him on the side of multinaturalism of biodiversity and eco-constructivism; but his promotion (using Deleuze as an intermediary) of difference against diversity aligns him closer with the multinaturalism of differentiation. Jamie Lorimer, *Wildlife in the Anthropocene: Conservation after Nature* (Minneapolis: University of Minnesota Press, 2015), 32–34.

57. See Alain Rey, ed., *Dictionnaire de la langue française* (Paris: Le Robert, 1992), s.v. "construire."

58. Aldo Leopold, *A Sand County Almanac and Sketches Here and There*, with illustrations by Charles W. Schwarz (Oxford: Oxford University Press, 1968), 132.

59. Bernard Charbonneau, "Le sentiment de la nature, force révolutionnaire," in *Nous sommes des révolutionnaires malgré nous*, by Bernard Charbonneau and Jacques Ellul (Paris: Seuil, 2014), 119–92.

60. Mumford, *Myth of the Machine*, 235.

61. [See Gilles Deleuze and Félix Guattari's *A Thousand Plateaus* (Minneapolis: University of Minnesota Press, 1987) for more on this concept of the full body.]

62. For more on the "Intrusion of Gaia," see Isabelle Stengers, *In Catastrophic Times: Resisting the Coming Barbarism*, trans. Andrew Goffey (London: Open Humanities Press, 2015), and *Une autre science est possible!* (Paris: La Découverte, 2013), 116–17. We would need a Melanie Klein in order to analyze the relation and engagement that the eco-constructivists have with Gaia.

63. Latour, *Facing Gaia*, 2, 242. As we will see later on in the present work, Latour is not afraid of our technological monsters, he pleads for us to love them; but one thing does frighten him: Gaia. One of the goals of our present work is to show how this fear is an unexamined effect of this love.

THE COPENHAGEN CHIASM

1. "Low Targets, Goals Dropped: Copenhagen Ends in Failure," *The Guardian*, December 18, 2009.

2. Concerning all of these examples, see "Copenhagen: Geoengineering's Big Break?," *Mother Jones*, December 14, 2009, as well Clive Hamilton, *Earthmasters: The Dawn of the Age of Climate Engineering* (New Haven, Conn.: Yale University Press, 2013), 14–16.

3. See Edwin Zaccai, "L'échec de Copenhague en perspective," *Esprit* 2 (2010): 6.

1. THE SCREEN OF GEOENGINEERING

1. John von Neumann, "Can We Survive Technology?," in *The Fabulous Future: America in 1980* (New York: E. P. Dutton & Company, 1955), 40.

2. Roger Revelle et al., "Atmospheric Carbon Dioxide," *Restoring the Quality of Our Environment*, Report of the Environmental Pollution Panel, President's Science

Advisory Committee, The White House, November 1965, Appendix Y4, 127. Cited by Naomi Klein, *This Changes Everything: Capitalism vs. the Climate* (New York: Simon & Schuster, 2014), 261.

3. Paul Crutzen, "Albedo Enhancement by Stratospheric Sulfur Injections: A Contribution to Resolve a Policy Dilemma?," *Climatic Change* 77 (2006): 216.

4. Klein, *This Changes Everything*, 262. For the video on how to "Save the Artic" thanks to this stratospheric shield, see IntellectualVentures, "The Stratoshield 'Hose to the Sky' Could Reverse Global Warming," YouTube video, 3:52, October 26, 2009, https://www.youtube.com/watch?v=JrimZzgqwdo. The project is described in Intellectual Ventures Laboratory, "The Stratospheric Shield," http://www.intellectualventureslab.com/assets_lab/Stratoshield-white-paper-300dpi.pdf.

5. Klein, *This Changes Everything*, 277.

6. See the conference, Fighting Fire with Fire: Climate Modification and Ethics in the Anthropocene (Sydney University and University of New South Wales, July 29–30, 2014), which we will speak about later on.

7. "What Is Geoengineering?," Oxford Geoengineering Programme, http://www.geoengineering.ox.ac.uk/what-is-geoengineering/what-is-geoengineering/.

8. Clive Hamilton, *Earthmasters: The Dawn of the Age of Climate Engineering* (New Haven, Conn.: Yale University Press, 2013), 172.

9. Crutzen, "Albedo Enhancement by Stratospheric Sulfur Injections," 217.

10. Martin Rees quoted in Alok Jha, "Astronomer Royal Calls for 'Plan B' to Prevent Runaway Climate Change," *The Guardian*, September 11, 2013. In regard to this argument for buying us some time, see Hamilton, *Earthmasters*, 159.

11. For a description of the conference, see "Fighting Fire with Fire: Climate Modification and Ethics in the Anthropocene," UNSW Sydney School of Humanities and Languages, https://hal.arts.unsw.edu.au/media/HALFile/1_Program__fire_with_fire_workshop.pdf.

12. One also speaks of a *nonlinear system* in terms of when consequences, amplified by positive retroactions, are *out of proportion* in relation to the causes that generated them. This disjunction between an initial change with a weak scope and a dramatic consequence is linked to the tipping points, these famous moments linked to the passage from one state in a system to a completely different system from the first. *The Day after Tomorrow* (dir. Roland Emmerich, 2004) is exemplary of such a tipping point in the climate system: First there is the slowing down of the Gulf Stream because of the influx of water in the Atlantic resulting from the melting of glaciers; after a certain tipping point, the Gulf Stream completely stops flowing, leading to a tipping point in the climate system itself—namely, a sudden, abrupt change in the system. As Tim Flannery perfectly shows, there is nothing about this scenario that is scientifically impossible, even if the climatologists do not consider it as an imminent danger; in any case, the chronology of an event of this kind is largely compressed within this fiction. Tim Flannery, *The Weather Makers: How Man Is Changing the Climate and What It Means for Life on Earth* (New York: Atlantic Monthly Press, 2005), 190–96.

13. Frederick Nebeker, "The Recognition of Limits of Weather Prediction," chap. 13 in *Calculating the Weather: Meteorology in the 20th Century* (New Brunswick, N.J.: Academic Press, 1995), 188–94.

14. Alan Robock, Luke Oman, and Georgiy L. Stenchikov, "Regional Climate Responses to Geoengineering with Tropical and Arctic SO_2 Injections," *Journal of Geophysical Research* (August 16, 2008), cited in Klein, *This Changes Everything*, 270.

15. Hamilton, *Earthmasters*, 173–77.

16. James Rodger Fleming, *Fixing the Sky* (New York: Columbia University Press, 2010), 8.

17. Fleming, *Fixing the Sky*, 8.

18. Regarding the dangers of geoengineering as "pure adaptation," see Dominique Bourg and Gerald Hess, "La géo-ingénierie: réduction, adaptation et scénario du désespoir," *Natures Sciences Sociétés* 18, no. 3 (2010): 298–304.

2. THE MIRROR OF THE ANTHROPOCENE

1. Jean-François Lyotard, *The Postmodern Explained* (Minneapolis: University of Minnesota Press, 1992).

2. See *The Anthropocene Review* (http://anr.sagepub.com).

3. Paul Crutzen and Eugene F. Stoermer, "The 'Anthropocene,'" *Global Change Newsletter* 41 (2000): 18.

4. 1. In fact, Watt did not "invent" the steam engine (whose initial so-called atmospheric model was invented by Denis Papin in 1690) but perfected a pre-existing model created by Thomas Newcomen. Watt's engine uses compressed (and not just atmospheric) vapor; the engine becomes autonomous and breaks away from the environment (the engine makes itself anatural [s'anaturalise]). 2. 1784 appears to correspond to the date of the patent that Watt placed for a steam locomotive.

5. Regarding the French term, *vérole* [which can be translated as a plague or sexually transmitted disease], being applied to humanity, see Dave Foreman and Murray Bookchin, *Quelle écologie radicale?* (Lyons: Atelier de création libertaire/Silence, 1994), 31.

6. William F. Ruddiman, *Plows, Plagues and Petroleum: How Humans Took Control of Climate* (2005; repr. Princeton, N.J.: Princeton University Press, 2010), 5. See as well the way in which Patrick V. Kirch, anthropologist and archaeologist, puts forth the hypothesis having the beginning of the transformation of the environment already beginning with the mastery of fire and, more specifically, insisting on the development of agriculture. Patrick V. Kirch, "Archaeology and Global Chance: The Holocene Record," *Annual Review of Environment and Resources* 30 (2005), 409–40.

7. Ruddiman, *Plows, Plagues and Petroleum*, 95–105.

8. Christophe Bonneuil and Jean-Baptiste Fressoz, *The Shock of the Anthropocene*, trans. David Fernbach (New York: Verso, 2016), 14.

9. Such a claim would do nothing but comfort the position held by the likes of a Paul Shepard, who saw within agriculture a veritable catastrophe that produced "the confusion that has plagued mankind for ten millennia" between "what is made by him

and what is not." Paul Shepard, *The Tender Carnivore and the Sacred Game* (New York: Scribner's and Sons, 1973), 274.

10. Ruddiman, *Plows, Plagues, and Petroleum*, 184.

11. Ruddiman, 179–80.

12. Ruddiman, 184.

13. Will Steffen, Wendy Broadgate, Lisa Deutsch, Owen Gaffney, and Cornelia Ludwig, "The Trajectory of the Anthropocene: The Great Acceleration," *The Anthropocene Review* (2015): 1–18. See as well Zalasiewicz et al., "When Did the Anthropocene Begin? A Mid-twentieth Century Boundary Level Is Stratigraphically Optimal," *Quarterly International* (2014): http://dx.doi.org/10.1016/j.quaint.2014.11.045.

14. We should note, in passing, that dispersal of fertilizer all over the planet smothers the fauna in the lakes and estuaries, contaminating the subterranean water tables and contributing to the warming of the planet.

15. Paul Crutzen and Christian Schwägerl, "Living in the Anthropocene: Toward a New Global Ethos," *Yale Environment 360*, January 24, 2011, https://e360.yale.edu/features/living_in_the_anthropocene_toward_a_new_global_ethos.

16. Simon L. Lewis and Mark A. Maslin, "Defining the Anthropocene," *Nature* 519 (2015): 175. Also see Dana Luciano, "The Inhuman Anthropocene," *Avidly*, March 2015.

17. "Welcome to the Anthropocene," video, 3:37, Welcome to the Anthropocene, www.anthropocene.info/index.php.

18. Concerning the question of modern science, see Edmund Husserl, *The Crisis of the European Sciences and Transcendental Phenomenology*, trans. David Carr (Evanston, Ill.: Northwestern University Press, 1970).

19. See Karl Marx, *Early Writings*, trans. Rodney Livingstone and Gregor Benton (London: Penguin Books/New Left Review, 1974), 329.

20. Marx, *Early Writings*, 329.

21. Sebastian Vincent Grevsmühl, *La Terre vue d'en haut. L'invention de l'environnement global* (Paris: Seuil, 2014), 219.

22. Benjamin Lazier, "Earthrise; Or the Globalization of the World Picture," *The American Historical Review* 116 (2011): 604. For more about *Heimatlosigeit* and "planetary man," see Michel Haar, *Heidegger and the Essence of Man* (Albany: State University of New York Press, 1993), 172. Concerning the terms *Heim* and *Heimat*, see Jean-Joseph Goux, "L'oubli de Hestia," *Langages* 85 (1987): 55.

23. Lazier, "Earthrise," 604.

24. Friedrich Nietzsche, *The Gay Science*, trans. Walter Kaufman (New York: Vintage Books, 1974), 181.

25. Nietzsche, *Gay Science*, 181.

26. Richard Buckminster Fuller, *Operating Manual for Spaceship Earth* (Baden, Switzerland: Lars Müller, 2008), 56.

27. The word *planet* also has the meaning of a traveler or vagabond. Alain Rey, *Dictionnaire historique de la langue française* (Paris: Dictionnaires Le Robert, 2011), loc. 239,136–239,139, Kindle.

28. See Alexandre Koyré, *From the Closed World to the Infinite Universe* (Baltimore, Md.: Johns Hopkins University Press, 1968).

29. Friedrich Nietzsche, *The Will to Power*, trans. Walter Kaufmann and R. J. Hollingdale (New York: Vintage Books, 1968), 8.

30. Fuller, *Operating Manual for Spaceship Earth*, 62.

31. Norbert Wiener, *The Human Use of Human Beings: Cybernetics and Society* (London: Free Association Books, 1989), 15.

32. Tim Ingold, "Humanity and Animality," in *Companion Encyclopedia of Anthropology: Humanity, Culture and Social Life*, ed. Tim Ingold (London: Routledge, 1994), 15.

33. Ingold, "Humanity and Animality," 17–18.

34. Ingold, 21.

35. Dipesh Chakrabarty, "The Climate of History: Four Theses," *Critical Inquiry* 35, no. 2 (Winter 2009): 197–222.

36. Concerning this point, see Jean-Jacques Kupiec and Pierre Sonigo, *Ni Dieu ni gene. Pour une autre théorie de l'hérédité* (Paris: Seuil, 2003). See also Frédéric Neyrat, *Homo labyrinthus. Humanisme, antihumanisme, posthumanisme* (Paris: Dehors, 2015), 78–85.

37. Chakrabarty, "Climate of History."

38. Dipesh Chakrabarty, "Postcolonial Studies and the Challenge of Climate Change," *New Literary History* 43, no. 1 (2012): 12–14. We should also mention Chakrabarty's analyses of postcolonial criticism—and the translation of this criticism by the thinkers of environmental justice—of "the human" in "Quelques failles dans la pensée sur le changement climatique," in *De l'univers clos au monde infini* (Paris: Dehors, 2014), 123–35.

3. TERRAFORMING: RECONSTRUCTING THE EARTH, RECREATING LIFE

1. Will Stewart [Jack Williamson], "Collision Orbit," *Astounding Science Fiction*, July 1942.

2. Carl Sagan, *Pale Blue Dot: A Vision of the Human Future in Space* (New York: Ballantine Books, 1997), xvi.

3. Sagan, *Pale Blue Dot*, xx.

4. Sagan, 339.

5. Sagan, 283.

6. Melville quoted in Sagan, xv.

7. William Cronon, "The Trouble with Wilderness; or Getting Back to the Wrong Nature," in *Uncommon Ground: Rethinking the Human Place in Nature*, ed. William Cronon (New York: W. W. Norton & Co., 1995), 85.

8. Frederick Jackson Turner, *History, Frontier, and Section* (Albuquerque: University of New Mexico Press, 1993).

9. Howard E. McCurdy, *Space and the American Imagination* (Baltimore: Johns Hopkins University Press, 2011), our summary of his discussion on 155–63.

10. McCurdy, *Space and the American Imagination*, 167. The illustration of the cosmic ark is on p. 169.

11. McCurdy, *Space and the American Imagination*, 171.

12. National Commission on Space, *Pioneering the Space Frontier: The Report of the National Commission on Space—An Exciting Vision of Our Next Fifty Years in Space*, https://www.nasa.gov/pdf/383341main_60%20-%2020090814.5.The%20Report%20of%20the%20National%20Commission%20on%20Space.pdf.

13. McCurdy makes a reference to the claims made in 1995 by Rick Tumlinson, cofounder of the Space Frontier Foundation (McCurdy, *Space and the American Imagination*, 172). For more about this question, see Sebastian Vincent Grevsmühl, "L'Astrofuturisme : colonies spatiales et laboratoires sociaux," in *La Terre vue d'en haut* (Paris: Seuil, 2014), 40–46.

14. Elizabeth Kolbert, "Project Exodus: What's behind the Dream of Colonizing Mars?," *New Yorker*, June 1, 2015, http://www.newyorker.com/magazine/2015/06/01/project-exodus-critic-at-large-kolbert.

15. See Jonathan Amos, "Obama Cancels Moon Return Project," *BBC News*, February 1, 2010, http://news.bbc.co.uk/2/hi/science/nature/8489097.stm.

16. Chris Impey, *Beyond: Our Future in Space* (New York: W. W. Norton, 2015), 65–66. But Impey maintains, with a relentless optimism, that a new era of space exploration is underway thanks to private enterprise (81–98).

17. "The End of the Space Age: Inner Space Is Useful. Outer Space Is History," *Economist*, June 30, 2011, http://www.economist.com/node/18897425.

18. In other words, the abandonment of the space program is not, as Matthew D. Tribbe claims, the effect of a reconsideration of progress and rationalism for the benefit of a "neo-romantic turn" but rather a reinvestment of technological rationality in regard to the Earth (Matthew D. Tribbe, *No Requiem for the Space Age: The Apollo Moon Landings and American Culture* [Oxford: Oxford University Press, 2014], 3–23). Is this not exactly what someone like Stewart Brand bears witness to, moving from a counterculture that fell in love with the Earth as seen from above and that has eventually led to a cyberculture of nuclear defense, GMOs, and biotechnologies? See Fred Turner, *From Counterculture to Cyberculture: Stewart Brand, the Whole Earth Network, and the Rise of Digital Utopianism* (Chicago: University of Chicago Press, 2006).

19. The forced digitalization of the 1990s, which was translated into the collective imaginary by the famous film *The Matrix* (dir. Lilly and Lana Wachowski, 1999), and the fear of a submersion into virtual spaces will have been one condition of possibility for a Reversal of the Frontier: The terrestrial globe has been sufficiently modeled in order to be materially transformed according to a planetary program of the geo-colonizers of the Earth. In this sense, we go from the cyborg years to the years of simulacra-nature.

20. "Purpose: Why We Go," Virgin Atlantic, https://www.virgingalactic.com/purpose/.

21. "Vision: Where We're Heading," Virgin Galactic, https://www.virgingalactic.com/vision/.

22. Rory Carroll, "Elon Musk's Mission to Mars," *Guardian*, July 17, 2013, https://www.theguardian.com/technology/2013/jul/17/elon-musk-mission-mars-spacex.

23. Kolbert, "Project Exodus."

24. "Le tourisme spatial cloué à Terre," *Le Monde*, November 4, 2014; "Un nouvel accident affaiblit le modèle SpaceX d'Elon Musk," *Le Monde*, September 2, 2016.

25. Kolbert, "Project Exodus."

26. "Project Persephone: British Scientists Building 'Living Space Ark' to Save Humanity," *Huffington Post UK*, April 28, 2014. For more about these questions, see the work of Clive Hamilton, "Dreams of a Fallen Civilisation: Why There Is No Escaping the Blue Planet," *ABC Religion and Ethics*, August 21, 2015.

27. "Paradis extrasolaires. Il existe des terres plus vivables que la nôtre," *Science et Vie*, July 2015, 57–70.

28. For more on these questions, see Sagan, *Pale Blue Dot*, 285.

29. James E. Lovelock, "A Physical Basis for Life Detection Experiments," *Nature* 207, no. 997 (1965): 568–70.

30. Concerning the relation between the Gaia hypothesis and terraforming, see, in particular, the chapter, "Greenhouse and Daisies," in *The Greening of Mars*, by James Lovelock and Michael Allaby (New York: St. Martin's/Marek, 1984), 91–108. You can also see syntheses of this question on pages 9 and 65.

31. Bernadette Bensaude-Vincent and Dorothée Benoit-Browaeys, *Fabriquer la vie. Où va la biologie de synthèse?* (Paris: Seuil, 2011), 65–69.

32. Kent H. Redford, William Adams, and Georgina M. Mace, "Synthetic Biology and Conservation of Nature: Wicked Problems and Wicked Solutions," *PLoS Biology* 11, no. 4 (2013).

33. Pascale Mollier, "Réflexion : la biologie de synthèse a besoin des sciences sociales," INRA Science and Impact, October 10, 2014, http://www.inra.fr/Chercheurs -etudiants/Economie-et-sciences-sociales/Tous-les-dossiers/biologie-de-synthese-et -sciences-sociales/Definir-la-biologie-de-synthese-un-premier-enjeu-de-debat.

34. Bensaude-Vincent and Benoit-Browaeys, *Fabriquer la vie*, 44–45.

35. Redford, Adams, and Mace, "Synthetic Biology and Conservation of Nature."

36. See the work of Vincent Devictor in *Nature en crise* (Paris: Seuil, 2015), 244–47. For Peter Kareiva, conservation has always led to instances of injustice (defending animals to the detriment of humans). It becomes even more impossible within the Anthropocene, since nothing is no longer truly wild. We must, therefore, make use of beneficial technologies and stop scolding capitalism. Peter Kareiva, Sean Watts, Robert McDonald, and Timothy M. Boucher, "Domesticated Nature: Shaping Landscapes and Ecosystems for Human Welfare," *Science* 316, no. 5833 (2007): 1866–69; Peter Kareiva, Michelle Marvier, and Robert Lalasz, "Conservation in the Anthropocene: Beyond Solitude and Fragility," *Breakthrough Journal* (2012).

37. Stewart Brand, who also wrote the "Ecomodernist Manifesto," is a fan of de-extinction. See his Ted Talk, "The Dawn of De-extinction: Are You Ready?," TED, March 13, 2013, video, 18:24, https://www.youtube.com/watch?v=XKc9MJDeqjo.

38. For more on this subject, see Jamie Lorimer, *Wildlife in the Anthropocene: Conservation after Nature* (Minneapolis: University of Minnesota Press, 2015), 97–100, and Elizabeth Kolbert, "Recall of the Wild: The Quest to Engineer a World before Humans," *New Yorker*, December 2012, 50–60. For a more general discussion on the topic,

see Steve Carver, "(Re)creating Wilderness: Rewilding and Habitat Restoration," in *The Routledge Companion to Landscape Studies*, eds. Peter Howard, Ian H. Thompson, and Emma Waterton (London: Routledge, 2013), 383–94.

39. "Will Dead Species Live Again?," interview with Stanley A. Temple, *Grow: Wisconsin's Magazine for the Life Sciences*, November 20, 2013.

40. "Will Dead Species Live Again?"

41. Redford, Adams, and Mace, "Synthetic Biology and Conservation of Nature."

42. Michel Foucault, "Right of Death and Power over Life," part 5 in *The History of Sexuality*, vol. 1, trans. Robert Hurley (New York: Vintage, 1990).

43. Catherine Larrère and Raphaël Larrère, *Penser et agir avec la nature* (Paris: La Découverte, 2015), 190, and, in a more general manner, the entire chapter 6, "Le démiurge et le pilote," 175–201.

44. Larrère and Larrère, *Penser et agir avec la nature*, 179.

45. Larrère and Larrère, 185.

46. Definition derived from the "Pilote" entry in Alain Rey, *Dictionnaire historique de la langue française* (Paris: Dictionnaires Le Robert, 2011), loc. 239,136–239,139, Kindle.

4. THE LOGIC OF GEOPOWER: POWER, MANAGEMENT, AND EARTH STEWARDSHIP

1. David Keith is a specialist in climate sciences and energy technologies as well as being an entrepreneur involved in public policy.

2. Quoted by Clive Hamilton, *Earthmasters: The Dawn of the Age of Climate Engineering* (New Haven, Conn.: Yale University Press, 2013), 73.

3. Hamilton, *Earthmasters*, 135.

4. F. Stuart Chapin, III et al., "Earth Stewardship: Science for Action to Sustain the Human-Earth System," *Ecosphere* 2, no. 8 (2011): article 89.

5. Alain Rey, *Dictionnaire historique de la langue française* (Paris: Dictionnaires Le Robert, 2011), loc. 239,140–239141, Kindle, s.v. "Intendant."

6. Victor Galaz, "Geo-engineering, Governance, and Social-Ecological Systems: Critical Issues and Joint Research Needs," *Ecology and Society* 17, no. 1: 24. See http://dx .doi.org/10.5751/ES-04677-170124, where you can download the article, which we will cite several times.

7. Galaz, "Geo-engineering, Governance, and Social-Ecological Systems," 1.

8. Galaz, 6.

9. Laura Ogden, Nik Heynen, Ulrich Oslender, Paige West, Karim-Aly Kassam, and Paul Robbins, "Global Assemblages, Resilience, and Earth Stewardship in the Anthropocene," *Frontiers in Ecology and the Environment* 11 (2013): 342.

10. Ogden et al., "Global Assemblages, Resilience, and Earth Stewardship," 343.

11. Ogden et al., 345.

12. See "Peoples Agreement," World People's Conference on Climate Change and the Rights of Mother Earth, April 22, 2010, Cochabamba, Bolivia, http://pwccc.wordpress .com/support/.

13. Christophe Bonneuil and Jean-Baptiste Fressoz, *The Shock of the Anthropocene*, trans. David Fernbach (New York: Verso, 2016), xi.

14. Alain Gras, *Le Choix du feu* (Paris: Fayard, 2007).

15. Bonneuil and Fressoz, *Shock of the Anthropocene*, 170–97.

16. This is why, initially, for Christophe Bonneuil and Jean-Baptiste Fressoz, the Anthropocene was first and foremost an "Anglocene" (*Shock of the Anthropocene*, 116).

17. See the work of Andreas Malm and Alf Hornborg, "The Geology of Mankind? A Critique of the Anthropocene Narrative," *The Anthropocene Review* (2014), 64.

18. Malm and Hornborg, "Geology of Mankind?," 67. In an article published in 2015, Andreas Malm had a less neutral response: "Dehistoricizing, universalizing, eternalizing, and naturalizing a mode of production specific to a certain time and place—these are the classic strategies of ideological legitimation." Andreas Malm, "The Anthropocene Myth," *Jacobin*, March 30, 2015, https://www.jacobinmag.com/2015/03/anthropocene-capitalism-climate-change/.

19. Malm and Hornborg, "Geology of Mankind?," 66.

20. Malm and Hornborg, 66.

21. Jason W. Moore, introduction to *Anthropocene or Capitalocene? Nature, History, and the Crisis of Capitalism*, ed. Jason W. Moore (Oakland, Calif.: PM Press/Kairos, 2016), 6. See also "The Rise of Cheap Nature," by the same author in the same volume, 85.

22. Moore, "Rise of Cheap Nature," 91, 111.

23. Paul Crutzen and Christian Schwägerl, "Living in the Anthropocene: Toward a New Global Ethos," *Yale Environment 360*, January 24, 2011, https://e360.yale.edu/features/living_in_the_anthropocene_toward_a_new_global_ethos. For more on the "Promethean discourse" established around the idea that nature doesn't exist, and has therefore been reduced to a transformable and controllable "raw material," see John S. Dryzek, *The Politics of the Earth: Environmental Discourses* (Oxford: Oxford University Press, 2013), 59–61.

24. Regarding an analysis of the term cyborg, see Frédéric Neyrat, *Homo labyrinthus: humanisme, antihumanisme, posthumanisme* (Paris: Editions Dehors, 2015), 134–37.

25. In other words, the Earth's radical "alterity." Concerning this idea and the difference between the globe and the planet, see the work of Gayatri Chakravorty Spivak, *Death of a Discipline* (New York: Columbia University Press, 2005), 72–74.

26. Mark Lynas becomes an apologist for nuclear energy as well as GMOs. *The God Species: Saving the Planet in the Age of Humans* (Washington, D.C.: The National Geographic Society, 2011). For more from the same author, see *Nuclear 2.0: Why a Green Future Needs Nuclear Power* (Cambridge: UIT Cambridge, 2013).

27. Regarding the intertwining of environment and social difference, see Gordon Walker, *Environmental Justice: Concepts, Evidence and Politics* (New York: Routledge, 2012).

28. Giovanna Di Chiro, "Ramener l'écologie à la maison," in *De l'univers clos au monde infini* (Paris: Dehors, 2014), 211.

29. Giovanna Di Chiro, "Nature as Community: The Convergence of Environment and Social Justice," in *Uncommon Ground: Rethinking the Human Place in Nature*, ed. William Cronon (New York: W. W. Norton & Company, 1996), 300–1.

30. Di Chiro, "Ramener l'écologie à la maison," 211.

31. Rob Nixon, *Slow Violence and the Environmentalism of the Poor* (Cambridge, Mass.: Harvard University Press, 2011). See as well the pioneering work by Ramachandra Guha and Joan Martínez-Alier, *Varieties of Environmentalism: Essays North and South* (London: Earthscan, 1997).

32. Cited by Naomi Klein in *This Changes Everything: Capitalism vs. the Climate* (New York: Simon and Schuster, 2015), 314.

33. "Political Ecology Begins When We Say 'Black Lives Matter,'" *Compass*, 2015, http://midwestcompass.org/political-ecology-begins-when-we-say-black-lives-matter /#more-1159.

TURBULENCE, RESILIENCE, DISTANCE

1. Bruno Latour, *Reassembling the Social: An Introduction to Actor-Network-Theory* (Oxford: Oxford University Press, 2005), 46.

5. AN ECOLOGY OF RESILIENCE: THE POLITICAL ECONOMY OF TURBULENCE

1. Karl Marx and Friedrich Engels, *The Communist Manifesto*, 1847. [We have chosen to translate the citation from the author's reference to the French edition of *The Communist Manifesto* by Marx and Engels. The exact wording of his citation may not appear exactly the same in recent English translations. For a more recent version of the text, see the following: "Constant revolutionizing of production, uninterrupted disturbance of all social conditions, ever-lasting uncertainty and agitation distinguish the bourgeois epoch from earlier ones. All fixed, fast-frozen relations, with their train of ancient and venerable prejudices and opinions are swept away, all new-formed ones become antiquated before they can ossify. All that is solid melts into thin air, all that is holy profaned, and man is at last compelled to face the sober senses of his kind." Karl Marx and Friedrich Engels, *The Communist Manifesto* (New York: International Publishers Co., 2014), 7.]

2. Ilya Prigogine and Isabelle Stengers, *Nouvelle Alliance. Metamorphose de la Science* (Paris: Editions Gallimard, 1979), 40. For the English editions see Ilya Prigogine and Isabelle Stengers, *Order out of Chaos* (New York: Bantam Books, 1984), 7–8.

3. For an analysis of the propositions put forth by Prigogine and Stengers, and of the genealogy that, through their work, runs through Lucretius to Michel Serres, see Frédéric Neyrat, *Clinamen. Flux, absolu et loi spirale* (Alfortville, France: è®e, 2011), 151–76.

4. What should be referenced here are approaches in terms of self-organization, at least those that have included the concept of "noise" as that from which (self-)organization begins (for example, I'm thinking of the work done by Henri Atlan, *Entre le cristal et le fumé* [Paris: Seuil, 1979]).

5. Melinda Cooper, "Turbulent Worlds: Financial Markets and Environmental Crisis," *Theory, Culture, Society* 27 (2010): 167–90.

6. Cooper, "Turbulent Worlds," 167.

7. Here we have largely borrowed from the work of Jeremy Walker and Melinda Cooper, "Genealogies of Resilience: From Systems Ecology to the Political Economy of Crisis Adaptation," *Security Dialogue* 42 (2011): 143–60. See as well, Russell B. Gallagher, "Risk Management: A New Phase of Cost Control," *Harvard Business Review* 34 (September–October 1956): 75–86; Gerry Dickinson, "Enterprise Risk Management: Its Origins and Conceptual Foundation," *The Geneva Papers on Risk and Insurance* 26, no. 3 (July 2001): 360–66.

8. Walker and Cooper, "Genealogies of Resilience," 149–50.

9. Concerning the ideology of chaos—that is to say, the ontological pseudo-justification of a chaos produced by an economic politics—see Frederick Buell, *From Apocalypse to Way of Life: Environmental Crisis in the American Century* (New York: Routledge, 2004), 223–25. Buell speaks about a "chaos propaganda machine."

10. See the work of Walker and Cooper, "Genealogies of Resilience," and Frederick Steiner, Mark Simmons, Mark Gallagher, Janet Ranganathan, and Colin Robertson, "The Ecological Imperative for Environmental Design and Planning," *Frontiers in Ecology and the Environment* 11, no. 7 (2013): 357.

11. "About Us," Stockholm Resilience Center, http://www.stockholmresilience.org/about-us.html.

12. C. S. Holling, "Resilience and Stability of Ecological Systems," *Annual Review of Ecology and Systematics* 4 (1973): 1–23.

13. Frederic Clements, "Nature and Structure of the Climax," in *Foundations of Ecology* (Chicago: University of Chicago Press, 1991), 59–97; Donald Worster, *The Wealth of Nature: Environmental History and the Ecological Imagination* (Oxford: Oxford University Press, 1993), 160.

14. Holling, "Resilience and Stability of Ecological Systems," 7.

15. Steiner et al., "Ecological Imperative for Environmental Design and Planning," 357. For more on the question of the two forms of resilience, see Lance Gunderson, Crawford S. Holling, Lowell Pritchard Jr., and Garry D. Peterson, "Resilience," in *Encyclopedia of Global Environmental Change*, ed. Ted Munn, vol. 2, *The Earth System: Biological and Ecological Dimensions of Global Environmental Change* (Chichester, UK: John Wiley & Sons, 2002), 530, as well as Lance H. Gunderson and Crawford S. Holling, eds., *Panarchy: Understanding Transformations in Human and Natural Systems* (Washington, DC: Island Press, 2002), 27–28.

16. Carl Folke, Steve Carpenter, Thomas Elmqvist, Lance Gunderson, C. S. Holling, and Brian Walker, "Resilience and Sustainable Development: Building Adaptive Capacity in a World of Transformations," *A Journal of the Human Environment* 31, no. 5 (2002): 437–38.

17. Folke et al., "Resilience and Sustainable Development," 437.

18. Holling, "Resilience and Stability of Ecological Systems," 7.

19. Gunderson and Holling, *Panarchy*, 5.

20. Gunderson and Holling, 14.

21. Gunderson and Holling, 31. For a good synthesis of the notions of adaptive cycles, panarchy, and environmental stewardship, with some more concrete examples, see Raphaël Mathevet and François Bousquet, *Résilience et environment. Penser le changements socio-écologiques* (Paris: Buchet-Castel, 2014).

22. Gunderson and Holling, *Panarchy*, 64–68.

23. For more on the concept of "surprise," see Gunderson and Holling, *Panarchy*, 13–14, 241–60, 315–32.

24. Folke et al., "Resilience and Sustainable Development," 438–39.

25. Gunderson and Holling, *Panarchy*, 67.

26. Alain Rey, *Dictionnaire historique de la langue française* (Paris: Dictionnaires Le Robert, 2011), loc. 239,140–239,141, Kindle.

27. Stephanie LeMenager and Stephanie Foote, "Editors' Column," *Resilience: A Journal of the Environmental Humanities* 1, no. 1 (January 2014). We will read with great interest Gay Hawkins's article about the political possibility consisting of seeing resilience as that which allows for *recreating the common*, of *re*-inventing a common usage, of *calling back into existence* something that had disappeared (in this case, the public, noncommercial use of water fountains). Gay Hawkins, "Resilience: The Resurgence of Public Things," *Resilience: A Journal of the Environmental Humanities* 1, no. 1 (January 2014).

6. THE EXTRAPLANETARY ENVIRONMENT OF THE ECOMODERNISTS

1. For a quick but illuminating synthesis of this phenomena, see Keith Kloor's work, "The Great Schism in the Environmental Movement: Can Modern Greens Loosen Nature's Grip on Environmentalism?," *Slate*, December 12, 2012. More recently, the "Ecomodernist Manifesto" no longer calls so much for the pure and simple liquidation of the environmentalist movement than its relegation to an "aesthetic" and "spiritual" function—namely, a function that is *anything but that which would have a function* that would be economic or political (John Asafu-Adjaye et al., "An Ecomodernist Manifesto," April 2015, www.ecomodernism.org/manifesto-english/, 27). See the critique of this text by Clive Hamilton, "The Technofix Is In," *Earth Island Journal*, April 21, 2015, http://www.earthisland.org/journal/index.php/elist/eListRead/the_technofix_is_in/.

2. John S. Dryzek, *The Politics of the Earth* (Oxford: Oxford University Press, 1997), 170–83.

3. For more on a critique of the "green industrial revolution" and "ecological modernization," see John Bellamy Foster, *The Ecological Revolution: Making Peace with the Planet* (New York: Monthly Review Press, 2009), 14–22 (one will also find in Foster's work a critique of the theses put forth by Ted Nordhaus and Michael Shellenberger), as well as the very useful chapter devoted to the Jevons paradox (121–28). Regarding the analysis of the ecological modernization under Bill Clinton, see Frederick Buell, *From Apocalypse to Way of Life* (New York: Routledge, 2004), 34–60.

4. Ted Nordhaus and Michael Shellenberger, "On Becoming an Ecomodernist: A Positive Vision of Our Environmental Future," The Breakthrough Institute, September 4, 2014, http://thebreakthrough.org/index.php/voices/michael-shellenberger-and-ted-nordhaus/on-becoming-an-ecomodernist.

5. See Kloor, "Great Schism in the Environmental Movement." Mark Lynas—who advocates for nuclear energy and GMOs (see *The God Species: Saving the Planet in the Age of Humans* [Washington, DC: The National Geographic Institute, 2011] and *Nuclear 2.0:*

Why a Green Future Needs Nuclear Power [Cambridge: UIT Cambridge, Ltd., 2014])—is clearly an ecomodernist; of this we have no doubts at all.

6. See Ted Nordhaus and Michael Shellenberger, "Fracktivists for Global Warming: How Celebrity NIMBYism Turned Environmentalism against Natural Gas," The Breakthrough Institute, March 6, 2013, https://thebreakthrough.org/index.php/voices/michael-shellenberger-and-ted-nordhaus/fracktivists-for-global-warming.

7. See, for example, Pascal Bruckner, "Les verts voudraient passer les menottes à la planète," interview by Christophe Barbier, *L'Express*, October 4, 2011.

8. Michael Shellenberger and Ted Nordhaus, "The Death of Environmentalism: Global Warming Politics in a Post-environmental World," The Breakthrough Institute, 2004, http://www.thebreakthrough.org/images/Death_of_Environmentalism.pdf, 9, 12, 19. See the analyses of Déborah Danowski and Eduardo Viveiros de Castro, "L'arrêt de monde," in *De l'univers clos au monde infini* (Bellevaux, France: Editions Dehors, 2014), 258–61.

9. Ted Nordhaus and Michael Shellenberger, eds., *Love Your Monsters: Postenvironmentalism and the Anthropocene* (Washington, DC: The Breakthrough Institute, 2011).

10. Nordhaus and Shellenberger, *Love Your Monsters*.

11. Ben A. Minteer and Stephen J. Pyne, *After Preservation* (Chicago: University of Chicago Press, 2015), 5.

12. Nordhaus and Shellenberger, *Love Your Monsters*.

13. Nordhaus and Shellenberger, *Love Your Monsters*.

14. Erle Ellis, "Overpopulation Is Not the Problem," *New York Times*, September 13, 2013, http://www.nytimes.com/2013/09/14/opinion/overpopulation-is-not-the-problem.html.

15. Erle Ellis, "The Planet of No Return: Human Resilience on an Artificial Earth," *Breakthrough Journal* 2 (Winter 2012), http://thebreakthrough.org/index.php/journal/past-issues/issue-2/the-planet-of-no-return.

16. For more on the distinction between tools and machines, see the work of Lewis Mumford, *Technics and Civilization*, foreword by Langdon Winner (1934; repr., Chicago: University of Chicago Press, 2010), 9–12.

17. Ellis, "Overpopulation Is Not the Problem."

18. Shellenberger and Nordhaus, *Love Your Monsters*.

19. Shellenberger and Nordhaus, "Death of Environmentalism," 8.

20. Stewart Brand, *Whole Earth Discipline: An Ecopragmatist Manifesto* (New York: Viking, 2009).

21. Asafu-Adjaye et al., "An Ecomodernist Manifesto." For more on David Keith, see "Buffering the Sun: David Keith and the Question of Climate Engineering," *Harvard Magazine*, July–August 2013.

7. THE "POLITICAL ECOLOGY" OF BRUNO LATOUR: NO ENVIRONMENTS, NO LIMITS, NO MONSTERS (NOT EVEN FEAR)

1. Bruno Latour, *Reassembling the Social: An Introduction to Actor-Network-Theory* (Oxford: Oxford University Press, 2005). In an article titled "On Recalling ANT," Latour

shows that the flaws of the *actor-network-theory* do not in any way invalidate the method (Bruno Latour, "On Recalling ANT," in *Actor Network Theory and After,* ed. John Law and John Hassard [Oxford: Blackwell Publishers/The Sociological Review, 1999], 15–25). Concerning the notion of composition, see Bruno Latour, "An Attempt at a 'Compositionist Manifesto,'" *New Literary History* 41 (2010): 471–90, http://www.bruno-latour.fr/sites/default/files/120-NLH-finalpdf.pdf.

2. Bruno Latour, "From Realpolitik to Dingpolitik—or How to Make Things Public," in *Making Things Public: Atmospheres of Democracy* (Cambridge, Mass.: MIT Press, 2005), 6.

3. Latour, "Attempt at a 'Compositionist Manifesto,'" 474–75. The call to no longer do critique—to no longer deconstruct, to no longer debunk, etc.—always runs the risk of veiling an area of thought that we would like to avoid from truly examining.

4. An interested reader can also find this article online here: Bruno Latour, "Love Your Monsters: Why We Must Care for Our Technologies as We Do Our Children," *Breakthrough Journal* 2 (Winter 2012), http://thebreakthrough.org/index.php/journal/past-issues/issue-2/love-your-monsters.

5. See Bruno Latour, *Aramis, or the Love of Technology* (Cambridge, Mass.: Harvard University Press, 1996), 83, 227, 248–49, 280.

6. Bruno Latour, *We Have Never Been Modern,* trans. Catherine Porter (Cambridge, Mass.: Harvard University Press, 1993), 11–12, 88.

7. Latour, *We Have Never Been Modern,* 106–7.

8. For more on this point see Michel Serres, *The Natural Contract* (Ann Arbor: University of Michigan Press, 1995), 33, and Ulrich Beck, *Risk Society: Towards a New Modernity* (London: Sage, 1992), 80–82.

9. Latour, "Love Your Monsters."

10. Latour, "Love Your Monsters."

11. Hans Jonas, *The Imperative of Responsibility: In Search of an Ethics for the Technological Age* (Chicago: University of Chicago Press, 1985).

12. Latour, "Love Your Monsters."

13. Bruno Latour, "'It's Development, Stupid!' or: How to Modernize Modernization," 10. This longer version, that has still yet to be officially published in its entirety, of "Love Your Monsters," can be found here: http://www.bruno-latour.fr/sites/default/files/107-NORDHAUS&SHELLENBERGER.pdf. The title is a reference to the saying that Bill Clinton used during his presidential campaign in order to discredit his opponent ("It's the economy, stupid!"). The signifier *development* is substituted here with that of *the economy* of which it is a synecdoche.

14. Bruno Latour, *Politics of Nature: How to Bring the Sciences into Democracy,* trans. Catherine Porter (Cambridge, Mass.: Harvard University Press, 2004), 26.

15. Latour, "'It's Development, Stupid!'"

16. Bruno Latour, *An Inquiry into Modes of Existence,* trans. Catherine Porter (Cambridge, Mass.: Harvard University Press, 2013), 8.

17. Bruno Latour, "Fifty Shades of Green," presentation at the Breakthrough Dialogue, Sausalito, California, June 2015.

18. Will Boisvert, "The Left vs. the Climate: Why Progressives Should Reject Naomi Klein's Pastoral Fantasy—and Embrace Our High-Energy Planet," Breakthrough Institute, September 18, 2014, https://thebreakthrough.org/index.php/programs/energy -and-climate/the-left-vs.-the-climate.

19. John von Neumann, "Can We Survive Technology?," *The Neumann Compendium*, ed. F. Bródy and T. Vámos (River Edge, N.J.: World Scientific Publishing, 1995), 670–71.

20. A syndrome that Lewis Mumford refers to as "the duty to invent" a "compulsion" that becomes active in the seventeenth century. Lewis Mumford, *Technics and Civilization*, with a foreword by Langdon Winner (1934; repr. Chicago: University of Chicago Press, 2010), 52–53.

21. Günther Anders, *The Obsolescence of Man*, vol. 2, *On the Destruction of Life in the Epoch of the Third Industrial Revolution*, posted by Alias Recluse, January 6, 2015, libcom .org, https://libcom.org/files/ObsolescenceofManVol%20IIGunther%20Anders.pdf.

22. Latour, "'It's Development, Stupid!,'" 12.

23. Francis Bacon, *The New Atlantis* (Ocean Shores, Wash.: Watchmaker Publishing, 2010), 42.

24. Latour, "'It's Development Stupid!'"

25. Latour, "Love Your Monsters."

26. Latour, "Love Your Monsters."

27. For more on all these ideas see Jean-Pierre Dupuy, *Pour un catastrophisme éclairé. Quand l'impossible est certain* (Paris: Seuil, 2002).

28. Bernard Stiegler, *Technics and Time*, vol. 1, *The Fault of Epimetheus*, trans. Richard Beardsworth and George Collins (Stanford, Calif.: Meridian Crossing, Stanford University Press, 1998), 183–203.

29. Mary Shelley, *Frankenstein; or The Modern Prometheus* (New York: Pocket Books, 2010), 113.

30. Shelley, *Frankenstein*, 34.

31. Shelley, 52.

32. Shelley, 44.

33. Shelley, 44.

34. René Descartes, *Discourse on Method*, 4th ed., trans. Donald A. Cross (Indianapolis: Hackett Publishing Company, 1998), 35. See as well: "The maintenance of health has always been one of the principal aims of my studies," letter to the Marquis of Newcastle, October 1645, cited in "Les idées de Descartes sur le prolongement de la vie et le mécanisme du vieillissement," *Revue d'histoire des sciences et de leurs applications* 21, no. 4 (1968): 291. Descartes seems to have slowly but surely moved from a consistent hope of "maintaining life" to that of "not fearing death."

35. Zygmunt Bauman, *Liquid Life* (Cambridge: Polity Press, 2005), 93.

36. Mumford, *Technics and Civilization*, 40.

37. In no way should our remarks here be construed as contesting the "legitimacy of the Modern Age" (Hans Blumenberg, *The Legitimacy of the Modern Age*, trans. Robert M. Wallace [Cambridge, Mass.: MIT Press, 1983]). Rather, they should be viewed

as demonstrating how modernity constructed itself in a singular manner by way of prolonging certain premodern characteristics—a bit as if the premodern was the unconscious of modernity, its active repression. In other words, it's not about claiming an absolute rupture, or flawless continuity, between scientific premodernity and scientific modernity but a *rupture through the method* and *continuity by way of an unconscious desire.*

38. Émilie Hache, *Ce à quoi nous tenons. Propositions pour une écologie pragmatique* (Paris: La Découverte, 2011), 12.

39. Latour, "Love Your Monsters."

40. During the third part of our present book, we will analyze Latour's *Facing Gaia: Eight Lectures on the New Climatic Regime,* where Latour, most notably, introduces the difference between humans and the Earthbound.

41. Latour, *Politics of Nature*, 41.

42. Latour, 159.

43. Bruno Latour, "The Promises of Constructivism," in *Chasing Technoscience: Matrix for Materiality*, ed. Don Idhe and Evan Sellinger (Bloomington: Indiana University Press, 2003), 31.

44. Latour, "Promises of Constructivism," 32.

45. Latour, 39.

46. Latour, 39.

47. Latour, *Politics of Nature*, 59.

48. See, for example: "Humans, dogs, oak trees, and tobacco are precisely on the same footing as glass bottles, pitchforks, windmills, comets, ice cubes, magnets, and atoms." Graham Harman, *Tool-Being: Heidegger and the Metaphysics of Objects* (Peru, Ill.: Carus Publishing Company, 2002), 2.

49. Alain Caillé, "Une politique de la nature sans politique. À propos de *Politiques de la nature* de Bruno Latour," *Revue du MAUSS* 17 (2001): 111.

50. Caillé, "Une politique de la nature sans politique," 113.

51. Latour, "Promises of Constructivism," 42.

52. "La catastrophe actuelle a été provoquée par l'imprudence des hommes," interview with Eisaku Sato, the former governor of Fukushima, in *Le Monde*, March 28, 2011; "Pour construire la centrale, Tepco avait raboté la falaise," *Le Monde*, July 12, 2011.

53. Nigel Clark, *Inhuman Nature: Sociable Life on a Dynamic Planet* (New York: Sage, 2010), 36–50.

8. ANATURALISM AND ITS GHOSTS

1. "An Ecomodernist Manifesto," http://www.ecomodernism.org/, 7.

2. Bruno Latour, *Politics of Nature: How to Bring Science into Nature*, trans. Catherine Porter (Cambridge, Mass.: Harvard University Press, 2004), 25–26.

3. Karl Marx and Friedrich Engels, *The German Ideology* (Amherst, N.Y.: Prometheus Books, 1998), 46.

4. Timothy Morton, *Ecology without Nature* (Cambridge, Mass.: Harvard University Press, 2009), 14.

5. Timothy Morton, "Ecology after Capitalism," *Polygraph* 22 (2010): 57. For Morton, our "ecological age" should force us to take into account that "everything is interconnected— interconnected in time too, then there's no 'over yonder,' and therefore no Nature"— an eco-constructivist slogan par excellence (40).

6. Slavoj Žižek, *In Defense of Lost Causes* (New York: Verso, 2008), 439.

7. Žižek, *In Defense of Lost Causes*, 444–45.

8. Michael Crichton, *State of Fear* (New York: HarperCollins, 2004).

9. Gary Snyder, "The Rediscovery of Turtle Island," in *Deep Ecology for the 21st Century*, ed. George Sessions (Boston: Shambhala, 1995), 454–62.

10. Arne Næss, *Ecology, Community, and Lifestyle: Outline of an Ecosophy*, trans. David Rothenberg (Cambridge: Cambridge University Press, 1993), 50.

11. Donald Worster, *The Wealth of Nature: Environmental History and the Ecological Imagination* (New York: Oxford University Press, 1993), 162.

12. Worster, *Wealth of Nature*, 164–65.

13. Paul Feyerabend, *Philosophy of Nature* (Cambridge: Polity, 2016), 28.

14. Alfred North Whitehead, *The Concept of Nature: Tarner Lectures* (Cambridge: Cambridge University Press, 2015), 12.

15. For more on the question of abstraction, see Alfred N. Whitehead, *Science and the Modern World* (New York: The Free Press, 1967), 54–55.

16. Aristotle, *Physics*, ed. David Bostock, trans. Robin Waterfield, Oxford World Classics (Oxford: Oxford University Press, 2008), 51. [While we have chosen to translate the following reference from the French translation of the author, for the interested reader, we have also provided a more recent translation that chooses to translate "art and techné" as "human skill." "And in general human skill either completes what nature is incapable of completing or imitates it."]

17. Cited by Carolyn Merchant referring to Richard Hooker in the opening pages of her book, *The Death of Nature: Women, Ecology, and the Scientific Revolution* (San Francisco: HarperOne, 1990), 6.

18. For more on Schiller's poem see Pierre Hadot, "Schiller's Gods of Greece," in *The Veil of Isis: An Essay on the History of the Idea of Nature*, trans. Michael Chase (Cambridge, Mass.: Belknap Press, 2008), 81–87.

19. See Martin Heidegger, "Séminaire du Thor," in *Questions III et IV* (Paris: Gallimard "Tel," 1990), 455–56.

20. Max Horkheimer and Theodor W. Adorno, *Dialectic of Enlightenment*, ed. Gunzelin Schmid Noerr, trans. Edmund Jephcott (Stanford, Calif.: Stanford University Press, 2007), 4.

21. Lewis Mumford, *Technics and Civilization*, foreword by Langdon Winner (1934; repr., Chicago: University of Chicago Press, 2010), 46–51.

22. René Descartes, *Meditations on First Philosophy*, ed. John Cottingham and Bernard Williams (Cambridge: Cambridge University Press, 1996), 20–21.

23. Whitehead, *Concept of Nature*, 18–64.

24. Merchant, *Death of Nature*, 2–41.

25. René Descartes, *The World and Other Writings*, ed. Stephen Gaukroger (Cambridge: Cambridge University Press, 1998), 25. Certainly Descartes's thoughts on Nature

are more complex, and Merleau-Ponty shows how Descartes takes into consideration the human body as that which can't be reduced to the status of an object (Maurice Merleau-Ponty, *Nature: Course Notes from the Collège de France* [Evanston, Ill.: University of Chicago Press, 2001], 18–19); but he nevertheless demonstrates that with Descartes, "the unified body is not the body itself but my body as thought by the soul" (20). Without any kind of autonomy.

26. Heraclitus, *The Art and Thought of Heraclitus: An Edition of the Fragments with Translation and Commentary*, ed. Charles H. Kahn (New York: Cambridge University Press, 1979), 105.

27. "From an active teacher and parent, she [that is, a nature personified as a woman] has become a mindless, submissive body." Merchant, *Death of Nature*, 190.

28. Razmig Keucheyan, *Nature Is a Battlefield: Towards a Political Ecology* (Cambridge: Polity, 2016), 84.

29. For more on cat bonds, see Keucheyan, *Nature Is a Battlefield*, 77.

30. Karl Marx and Friedrich Engels, *The Economic and Philosophic Manuscripts of 1844 and The Communist Manifesto*, trans. Martin Milligan (Amherst, N.Y.: Prometheus Books, 1988), 146. [There are a number of variants of this translation including: "Logic, is the currency of the mind," "Logic, is the money of the mind," and "Logic is the coin of the realm."]

31. Philippe Descola, *Beyond Nature and Culture*, trans. Janet Lloyd (Chicago: University of Chicago Press, 2013), 6, 14.

32. Descola, *Beyond Nature and Culture*, 4–5, 7.

33. Descola, 7.

34. Descola specifies that for the Achuar, "most insects and fish, grasses, mosses, and brackens, and pebbles and rivers remain outside the social sphere and outside the network of intersubjectivity." But, he clarifies, this is simply a "segment of the world" incomparably more restrained than what "we" understand as nature, and "that furthermore, 'nature' only has meaning when set in opposition to human works" (Descola, 7–8). A segment of the world is still a segment! This is what complicates the schema consisting of thinking that cultures express *either* a nature/culture dualism *or* an absence of dualism. And what if cultural reality was less binary that what Descola proposes?

35. Descola, 65.

36. Descola, xv, 173. We should also note that there is no mention of the work done by Carolyn Merchant, whereas even Descola speaks of the "mechanistic revolution."

37. Descola, 66.

38. "However, since Descartes, and above all Darwin, we have no hesitation in recognizing that the physical component of our humanity places us in a material continuum within which we do not appear to be unique creatures anymore significant than any other organized being" (Descola, 173). Far from being minoritarian or heretical, for Descola, "Darwin proposed the *canonical* version [our emphasis] of this incorporation of culture into nature in *The Descent of Man*" (Descola, 199).

39. Jean-Christophe Bailly, *The Animal Side*, trans. Catherine Porter (New York: Fordham University Press, 2011), 14–15.

40. Bill McKibben, *The End of Nature* (1989; repr., New York: Random House, 2006), 7. Richard White wrote an acerbic article about McKibben's books, criticizing the lack of theoretical depth and historical perspective, as well as his Emersonian vision of nature: "McKibben needs nature as an independent source of truth distinct from us. The disappearance of that separateness was the cri de cœur at the core of *The End of Nature*, and in his insistence on the separation of humans from a natural world, it is hard not to hear echoes of Walt Disney" (Richard White, "Bill McKibben's Emersonian Vision," *Raritan: A Quarterly Review* 31, no. 2 [2011], 113). Nevertheless, at the end of his article, Richard White recognizes the real political implications of McKibben in the environmental struggle. On Richard White's critique of McKibben's conception of nature, see Anna L. Peterson, *Being Human: Ethics, Environment, and Our Place in the World* (Berkeley: University of California Press, 2001), 67–68.

41. Paul Crutzen and Christian Schwägerl, "Living in the Anthropocene: Toward a New Global Ethos," *Yale Environment 360*, January 24, 2011, https://e360.yale.edu /features/living_in_the_anthropocene_toward_a_new_global_ethos.

42. Latour, *Politics of Nature*, 25, 28.

43. Latour, 33.

44. Latour, 254 (footnote 13 for pages 17–19).

45. Bruno Latour, *Facing Gaia: Eight Lectures on the New Climatic Regime*, trans. Catherine Porter (Cambridge: Polity, 2017), 144.

46. Catherine Larrère and Raphael Larrère, *Du bon usage de la nature. Pour une philosophie de l'environement* (Paris: Aubier, 1997), 11.

47. Edgar Morin, "La pensée écologisée. Pour une nouvelle conscience planétaire," *Le Monde diplomatique*, October 1989, 1. See the way in which Michael Löwy and Robert Sayre describe Romanticism as a transhistorical "collective mental structure" having as a common point the refusal of a "way of life in a capitalist society," of a disenchantment with the world, of a reification of the market, of cold rationalism and the destruction of social ties (Michael Löwy and Robert Sayre, *Révolte et mélancolie* [Paris: Payot, 1992]). We refer you as well to the volume "Politiques romantiques" that we edited for the revue *Multitudes* 55 (May 2014).

48. Merchant, *Death of Nature*, 293.

49. Morin, "La Pensée Ecologisée," 18–19.

50. For an analysis of *Meek's Cutoff*, see Frédéric Neyrat, "Le cinéma éco-apocalyptique. Anthropocène, Cosmophagie, Anthropophagie," *Communications* 96 (2015): 75–76.

51. For a good summary of the reasons for this anti-CO_2 campaign, see Bill McKibben, "Global Warming's Terrifying New Math: Three Simple Numbers That Add up to Global Catastrophe—and That Make Clear Who the Real Enemy Is," *Rolling Stone*, July 19, 2012.

9. THE TECHNOLOGICAL FERVOR OF ECO-CONSTRUCTIVISM

1. Bruno Latour, "Agency at the Time of the Anthropocene," *New Literary History* 45, no. 1 (Winter 2014): 1–18.

2. Naomi Klein, *This Changes Everything: Capitalism vs. the Climate* (New York: Simon & Schuster, 2015), 279.

3. Lance H. Gunderson and Crawford S. Holling, eds., *Panarchy: Understanding Transformations in Human and Natural Systems* (Washington, DC: Island Press, 2002), 67.

4. Bruno Latour, "A quoi nous-tenons vraiment, nous les Modernes?," *Le Nouvel Observateur*, May 18, 2014.

5. Pierre Gattaz, "N'orientons pas la France vers la décroissance," *Le Monde*, September 18, 2013.

6. For more on "French Theory," see François Cusset's *French Theory: How Foucault, Derrida, Deleuze, & Co. Transformed the Intellectual Life of the United States*, trans. Jeff Fort (Minneapolis: University of Minnesota Press, 2008).

7. Robin Mackay and Armen Avanessian, introduction to *#Accelerate: The Accelerationist Reader*, ed. Robin Mackay and Armen Avanessian (Falmouth, UK: Urbanomic, 2014), 23–46. Nick Land, *Fanged Noumena: Collected Writings 1987–2007* (Falmouth, UK/ Brooklyn, N.Y.: Sequence/Urbanomic, 2012). We should note that the term *accelerationist* was attributed to Nick Land by Benjamin Noys (*Malign Velocities: Capitalism and Accelerationism* [Winchester: Zero Books, 2014]). We also refer to the critique made by Déborah Danowski and Eduardo Viveiros de Castro in "L'arrêt de monde," in *De l'univers clos au monde infini*, ed. Émilie Hache (Paris: Dehors, 2014), 263–64.

8. Alex Williams and Nick Srnicek, "#Accelerate Manifesto for an Accelerationist Politics," Critical Legal Thinking, May 14, 2013, http://criticallegalthinking.com/2013 /05/14/accelerate-manifesto-for-an-accelerationist-politics/.

9. Williams and Srnicek, "#Accelerate Manifesto," point 5: "Accelerationists want to unleash latent productive forces. In this project, the material platform of neoliberalism does not need to be destroyed. It needs to be repurposed towards common ends. The existing infrastructure is not a capitalist stage to be smashed, but a springboard to launch towards post-capitalism."

10. Williams and Srnicek, point 21.

11. Ray Brassier, "Prometheanism and Its Critics," in Mackay Avanessian, *#Accelerate*, 485.

12. Brassier, "Prometheanism and Its Critics," 486–87.

13. Brassier, 487.

14. Williams and Srnicek, "#Accelerate Manifesto," point 22.

15. For more on the question of the posthuman within the transhumanist perspective, see Nick Bostrom, "Why I Want to Be a Posthuman When I Grow Up," in *Medical Enhancement and Posthumanity*, ed. Bert Gordijn and Ruth Chadwick (New York: Springer, 2008), 107–37. This position and the evolutionary theory that underpins it are critiqued in Frédéric Neyrat's *Homo labyrinthus: humanisme, antihumanisme, posthumanisme* (Paris: Dehors, 2015), 121–50.

16. André Gorz, *The Immaterial*, trans. Chris Turner (London: Seagull Books, 2010), 164.

17. See Ray Kurzweil, *The Singularity Is Near: When Humans Transcend Biology* (New York: Penguin, 2005). For more on this question, see Danowski and Viveiros de Castro, "L'arrêt de monde," 256–57.

18. One of the versions can be found here: "Transhumanist Declaration," Humanity+, http://humanityplus.org/philosophy/transhumanist-declaration/.

19. Nick Bostrom, "Existential Risk Prevention as Global Priority," *Global Policy* 4, no. 1 (2013): 15.

20. Bostrom, "Existential Risk Prevention as Global Priority," 16.

21. University of Cambridge Centre for the Study of Existential Risk, https://www.cser.ac.uk/about-us/. Here one will benefit from reading an article by one of its founders, Huw Price: "Cambridge, Cabs, and Copenhagen: My Route to Existential Risk," (*New York Times*, January 21, 2013). Price explains that the risks will perhaps come less from diabolical artificial intelligences than from stupid technologies—for example, from poorly constructed nanomachines "dumb optimizers," machines with rather simple goals (producing IKEA furniture, say) that figure out that they can improve their output astronomically by taking control of various resources on which we depend for our survival.

22. "The Rise of the Up-Wingers Part One: Steve Fuller on the Proactionary Principle, Environmentalism, and Interstellar Flight," Breakthrough Institute, October 21, 2014, http://thebreakthrough.org/index.php/issues/technology/the-rise-of-the-up-wingers. For more on Fuller's position in relation to transhumanism, see the interview: Steve Fuller, "Talking to the Future Humans," interview by Kevin Holmes, Vice, September 29, 2011, http://www.vice.com/en_uk/read/talking-to-the-future-humans-steve-fuller-transhumanism.

23. See Sahotra Sarkar, "Science v. Religion? Intelligent Design and the Problem of Evolution," review of *Science v. Religion? Intelligent Design and the Problem of Evolution*, by Steve Fuller, *Notre Dame Philosophical Review* (August 2008), https://ndpr.nd.edu/news/science-v-religion-intelligent-design-and-the-problem-of-evolution/.

24. Peter Sloterdijk, *La Domestication de l'être* (Paris: Mille et une nuits, 2000), 91.

25. Carl Folke, Steve Carpenter, Thomas Elmqvist, Lance Gunderson, Crawford S. Holling, and Brian Walker, "Resilience and Sustainable Development," *Ambio: A Journal of the Human Environment* 31, no. 5 (2002): 438–39.

26. Sloterdijk, *La Domestication de l'être*, 91.

27. André Gorz, *The Immaterial*, trans. Chris Turner (London: Seagull Books, 2010), 194–200.

28. André Gorz, *Ecologica*, trans. Chris Turner (London: Seagull Books, 2010), 9.

29. There is a "radical monopoly" when "an overefficient tool" "imposes compulsory consumption and, and thereby restricts personal autonomy. It constitutes a special kind of social control because it is enforced by means of the imposed consumption of a standard product that only large institutions can provide." Ivan Illich, *Tools for Conviviality* (1973; repr., London: Marion Boyars, 2001), 61–63.

30. The term *individuation* used here refers back to the early work by Gilbert Simondon, in particular *On the Mode of Existence of Technical Objects*, trans. Cecile Malaspina and John Rogove (Minneapolis: University of Minnesota Press, Univocal, 2017).

31. [A *Zadist* is a rather recent neologism referring to the French term *ZAD*, which is short for "Zone à defender" (a zone worth protecting). *Zadist* is, therefore, a recent entry into the French dictionary for those who participate in defending these territories.]

32. Williams and Srnicek, "#Accelerate Manifesto."

33. Isabelle Stengers, *Another Science Is Possible: Manifesto for a Slow Science*, trans. Stephen Muecke (London: Polity, 2018), 149.

34. Stengers, *Another Science Is Possible*, 81–82.

35. Donna Haraway, "Anthropocene, Capitalocene, Plantationocene, Chthulucene: Making Kin," *Environmental Studies* 6 (2015): 160, http://environmentalhumanities.org /arch/vol6/6.7.pdf.

36. Haraway, "Anthropocene, Capitalocene, Plantationocene, Chthulucene," 160–61.

37. Haraway, 161.

38. Haraway, 164n17.

39. Haraway, 161.

40. And not the "insistence of the possible." See Isabelle Stengers, "L'insistence du possible," in *Gestes Speculatifs*, ed. Didier Debaise and Isabelle Stengers (Paris: Presses du réel, 2015), 5–22.

41. Nigel Clark, *Inhuman Nature: Sociable Life on a Dynamic Planet*, Theory, Culture, and Society (Washington, DC: Sage Publications, 2010).

42. Clark, *Inhuman Nature*, xvi, 27–54.

OBJECT, SUBJECT, TRAJECT

1. While working on the American edition of the present book, we came across the illuminating analyses of Augustin Berque, who has also theorized the notions of the traject and "trajectivity" and to whom we owe a great amount of gratitude. For more on his work, we recommend Augustin Berque, *La mésologie, pourquoi et pour quoi faire?* (Nanterre, France: Presse Universitaire de Paris Ouest, 2014), concerning trajection and "trajective chains" [chaînes trajectives], 39–77.

10. NATURING NATURE AND NATURED NATURE

1. Olga Weijers, "Contribution à l'histoire des termes *'natura naturans'* et *'natura naturata'* jusqu'à Spinoza," *Vivarium* 16, no. 1 (1978): 70–80.

2. Aristotle, *Metaphysics*, in *Complete Works*, vol. 1 (Princeton, N.J.: Princeton University Press, 1984), 63 (Δ 4, 1014b16–1015a12).

3. For more on this point, see the studies conducted by Alexandre Koyré, *Études galiléennes* (Paris: Hermann, 1966), trans. by John Mepham as *Galileo Studies* (Atlantic Highlands, N.J.: Humanities Press, 1978).

4. Aristotle, *Physics*, trans. Robin Waterfield, Oxford World's Classics (New York: Oxford University Press, 2008), Book II, 193b12–20, 50.

5. Frédéric Manzini, *Spinoza, une lecture d'Aristote* (Paris: PUF, 2009), 242–43.

6. Spinoza, *Short Treatise on God, Man, and Human Welfare*, trans. Lydia Gillingham Robinson (Chicago: Open Court Publishing Co., 1909), 50.

7. [Spinoza, *Ethics*, trans. Edwin Curley (New York: Penguin Classics, 1996), 118 (IV, 4, Dem.). In this edition, Spinoza's famous expression, *Deus, sive Natura* is translated in its

common English equivalent "God, or Nature." However, we have chosen to maintain the Latin spelling referred to by the author in the original French version of his text.]

8. René Descartes, *Discourse on Method*, trans. Donald A. Cress, 4th ed. (Indianapolis, Ind.: Hackett Publishing, 1998), part 5, 23–33.

9. David Rothenberg and Arne Næss, *Is It Painful to Think? Conversations with Arne Næss* (Minneapolis: University of Minnesota Press, 1992), 85.

10. Rothenberg and Næss, *Is It Painful to Think?*, 92.

11. Spinoza, *Ethics*, 68.

12. Concerning the analogy between animals/clocks, see Descartes, *Discourse on Method*, 28.

13. Spinoza, *Ethics*, 120. Concerning the question of the dynamism of nature, see Manzini, *Spinoza*, 245, 271.

14. Spinoza, *Ethics*, 135. This critique is made by Genevieve Lloyd in "Spinoza's Environmental Ethics," *Inquiry* 23 (1980): 293–311. See as well Næss's response to this critique that appeared in the same issue of *Inquiry*: "Environmental Ethics and Spinoza's Ethics: Comments on Genevieve Lloyd's Article," 313–25.

15. Gilles Deleuze, *Spinoza: Practical Philosophy*, trans. Robert Hurley (San Francisco: City Lights, 1988), 122–30.

16. Gilles Deleuze, *Expressionism in Philosophy: Spinoza*, trans. Martin Joughin (New York: Zone Books, 1992), 345.

17. We are, of course, referring here to Spinoza's *Tractatus Theologico-Politicus*.

18. Deleuze, *Expression in Philosophy*, 172.

19. Maurice Merleau-Ponty, *Nature: Course Notes from the Collège de France*, trans. Robert Vallier, Studies in Existential Phenomenology (Evanston, Ill.: Northwestern University Press, 2003), 37.

20. Spinoza, *Ethics*, 10.

21. Georg Wilhelm Friedrich Hegel, *Lectures on the History of Philosophy*, vol. 3, *Medieval and Modern Philosophy*, trans. E. S. Haldane and Frances H. Simson (Lincoln: University of Nebraska Press, 1995), 281–82.

22. Alfred North Whitehead, *Process and Reality*, ed. David Ray Griffin and Donald W. Sherburne (New York: The Free Press, 1978), 81.

23. Pierre Macherey, *Hegel or Spinoza*, trans. Susan M. Ruddick (Minneapolis: University of Minnesota Press, 2011), 147, 201.

24. Martial Gueroult, cited by Macherey in *Hegel or Spinoza*, 180. For Deleuze, "Cartesianism was never the thinking of Spinoza; it was more like his rhetoric" (Deleuze, *Spinoza*, 8.). But Deleuze recognizes nevertheless that Spinoza conserves "the chief discovery of Cartesian mechanism": The powers of nature are actual and not hidden [occultes] (Deleuze, *Expressionism in Philosophy*, 228). As for Manzini, he speaks of Spinoza's "mitigated mechanism" using Aristotle as a "guardrail" against the excess of a mechanism "which he nevertheless profits from" (Manzini, *Spinoza, une lecture d'Aristote*, 272).

25. Deleuze, *Expressionism in Philosophy*, 270. On the naturalism "from Lucretius to Nietzsche" as a form of opposition to gods and as a form or pure affirmation, sensuality, and a recognition of the multiple, see Deleuze, *Logic of Sense*, ed. Constantin V.

Boundas, trans. Mark Lester with Charles Stivale (New York: Columbia University Press, 1990), 278–79.

26. Deleuze, *Spinoza*, 124.

27. Gilbert Simondon, *Individuation psychique et collective* (Paris: Aubier, 2007), 230.

28. Arne Næss, *Ecology, Community, and Lifestyle: Outline of an Ecosophy*, trans. David Rothenberg (1989; repr., New York: Cambridge University Press, 1990), 165.

29. Næss, *Ecology, Community, and Lifestyle*, 49.

30. Murray Bookchin, *Remaking Society* (Boston: South End Press, 1990), 7–40.

31. Bookchin, *Remaking Society*, 21.

32. Næss, *Ecology, Community, and Lifestyle*, 29.

33. See Karen Warren, *Ecological Feminism* (London: Routledge, 1994), 88–105.

34. Arne Næss implores us that "the uniqueness of humankind should not be underestimated" and to go against this "tendency" deep ecology has of "ridding itself" of the differences and that deep ecology "always underlines what we have in common with other life forms." Næss, *Ecology, Community and Lifestyle*, 169–71.

35. Deleuze, *Spinoza*, 128. See as well: "So, an animal, a thing, is never separable from its relations with the world. The interior is only a selected exterior, and the exterior, a projected interior" (125).

36. Gilles Deleuze and Félix Guattari, *Anti-Oedipus: Capitalism and Schizophrenia*, trans. Robert Hurley, Mark Seem, and Helen R. Lane (Minneapolis: University of Minnesota Press, 1983), 4. A page later on, the schizo is defined as *"Homo natura"* (5).

37. Deleuze and Guattari, *Anti-Oedipus*, 2.

38. Literally, *existence* means to stand (*sistere*) outside (*ex*).

39. Félix Guattari, *Qu'est-ce que l'écosophie?* (Paris: Nouvelle Editions Lignes, 2014), 533.

40. Guattari, *Qu'est-ce que l'écosophie?*, 73, 108, 65.

41. Guattari, 500, 113–15.

42. Guattari, 531.

43. Hannah Arendt, *The Human Condition*, 2nd ed. (Chicago: University of Chicago Press, 1998), 151, 309.

44. Günther Anders, L'Obsolescence de l'homme, vol. 2, *Sur la destruction de la vie à l'époque de la troisième révolution industrielle*, trans. Christophe David (Paris: Editions Fario, 2012), 21–22, 28, 68, 74–75.

45. Anders, *L'Obsolescence de l'homme*, 402.

11. THE REAL NATURE OF AN ECOLOGY OF SEPARATION

1. Barry Commoner, *The Closing Circle: Nature, Man, and Technology* (New York: Knopf, 1971), 39. For more on Næss's citation of Commoner, see "Spinoza and Ecology," *Philosophia* 7, no. 1 (March 1977): 44–54.

2. Commoner, *The Closing Circle*, 36–39.

3. Commoner, 41–42.

4. Jean-Paul Deléage, *Une histoire de l'écologie* (1991; repr., Paris: Seuil, "Points-Seuil," 1994), 19–80.

5. Deléage, *Une histoire de l'écologie*, 79.

6. John Muir, *My First Summer in the Sierra*, cited by Donald Worster, *The Wealth of Nature* (New York: Oxford University Press, 1993), 169.

7. Arne Næss, *Ecology, Community, and Lifestyle: Outline of an Ecosophy*, trans. David Rothenberg (1989; repr., New York: Cambridge University Press, 1990), 49, 56.

8. Isabelle Stengers, *Pour en finir avec la tolérance* (Paris: La Découverte/Les Empêcheurs de penser en rond, 1997), 74.

9. Stacy Alaimo, *Bodily Natures: Science, Environment, and the Material Self* (Bloomington: Indiana University Press, 2010), 2.

10. Graham Harman, *The Quadruple Object* (Alesford, UK: Zero Books, 2011).

11. Quentin Meillassoux, *After Finitude: An Essay on the Necessity of Contingency*, trans. Ray Brassier (London: Continuum, 2008).

12. See Freud's short essay, "Splitting of the Ego in the Process of Defence," in *The Standard Edition of The Complete Psychological Works of Sigmund Freud*, vol. 23, ed. James Strachey et al. (London: Hogarth Press and the Institute of Psychoanalysis, 1964), 271–78.

13. Heidegger uses the term *abyss* (*Abgrund*) to qualify the integral rupture between humans and animals. He claims whereas "man is master and servant of the world, in the sense of *'having'* world, . . . the animal is *poor in world*. Man is *world-forming*." Martin Heidegger, *Fundamental Concepts of Metaphysics: World, Finitude, Solitude*, Studies in Continental Thought (Bloomington: Indiana University Press, 2001), 177, 264.

14. [Saint Augustine, *Confessions*, III, 6, 11. Augustine's equation applies to God. I've tried to adhere to a literal translation of the French reference to the Latin version of Augustine's *Confessions*. For a more widely available English translation see R. S. Pine Coffin's translation of Saint Augustine's *Confessions*, "Yet you were deeper than my inmost understanding and higher than the topmost height that I could reach."] Saint Augustine, *Confessions*, trans. R.S. Pine Coffin (New York: Penguin Classics, 1961), 62.

15. Werner Heisenberg, *The Physicist's Conception of Nature* (London: Hutchinson & Co., 1958), 25.

16. Heisenberg, *The Physicist's Conception of Nature*, 24.

17. Heisenberg, 30.

18. Alfred N. Whitehead, *The Concept of Nature: Tarner Lectures Delivered in Trinity College November 1919* (London: Cambridge University Press, 1920), 52–55, 69, 170. *Process and Reality* will emphasize the difference between occasion (a process by which something becomes) and event (a collection of occasions). For more on this point, see Steven Shaviro, *Without Criteria: Kant, Deleuze, Whitehead, and Aesthetics* (Cambridge, Mass.: MIT Press, 2009), 18–19.

19. Alfred N. Whitehead, "The Romantic Reaction," in *Science and the Modern World* (New York: The Free Press, 1967), 75–94. See as well, the illuminating commentaries on Whitehead's *Philosophy in The Universe of Things: On Speculative Realism* (Minneapolis: University of Minnesota Press, 2014).

20. Whitehead, *Concept of Nature*, 4.

12. DENATURING NATURE

1. F. W. J. Schelling, *First Outline of a System of the Philosophy of Nature*, trans. Keith R. Peterson (Albany: SUNY Press, 2001), 202–3.

2. This is the "enigma of the world," Schelling adds (*Premiers Écrits*, Épiméthée [Paris: PUF, 1987], 180) in a question that takes on the impasses found in Spinoza. See as well: "The passage from the infinite to the finite constitutes precisely the *entire* problem of philosophy" (Le passage de l'infini au fini constitue précisément le problème de *toute* philosophie) (185).

3. Schelling, *First Outline of a System of the Philosophy of Nature*, 201, 204.

4. Schelling, 205.

5. Schelling, 213.

6. Gilles Deleuze and Félix Guattari, *Anti-Oedipus: Capitalism and Schizophrenia*, vol. 1, trans. Robert Hurley (Minneapolis: University of Minnesota Press, 1983), 10. A quasi-cause is an effect (which is produced) passing itself off as a cause.

7. Deleuze and Guattari, *Anti-Oedipus*, 262.

8. Schelling, *First Outline of a System of the Philosophy of Nature*, 215.

9. Pierre Hadot, *The Veil of Isis*, trans. Michael Chase (Cambridge, Mass.: Belknap Press, 2008), 8–9.

10. Martin Heidegger, cited by Marlène Zarader, *La Dette impensée. Heidegger et l'héritage hébraïque* (Paris: Seuil, 1990), 147.

11. Martin Heidegger, *Nietzsche II*, trans. David Farrell Krell (San Francisco: HarperOne, 1991).

12. Zarader, *La Dette impensée*, 149.

13. Alexandre Safran, *La Cabale*, cited in Marlène Zarader, *La Dette impensée*, 149.

14. F. W. J. Schelling, *The Ages of the World*, trans. Jason M. Wirth (Albany: SUNY Press, 2000), 16, 107.

15. Schelling, *Ages of the World*, 107.

16. Schelling, 30.

17. Alfred N. Whitehead, *Science and the Modern World* (New York: Free Press, 1997), 176.

18. To refer to a theory of the "ages of the world" that would be much more contemporary, would there be some sort of *big crunch* preceding every *big bang*?

19. Jean Paul Richter, *Flower, Fruit and Thorn Pieces* (Boston: James Munroe and Company, 1845), 336.

20. Gérard de Nerval, "*Christ at Gethsemane* (from *The Chimeras*)," trans. Henry Weinfield, *Literary Imagination* 8, no. 2 (2006): 229–31.

21. The death of God was clearly formulated by Hegel before Nietzsche: "God himself is dead." G. W. F. Hegel, *Faith and knowledge* (Philadelphia: University of Pennsylvania Press, 1988), 190.

22. G. W. F. Hegel, "Jena Lectures, 1805," in *Hegel and The Human Spirit: A Translation of the Jena Lectures*, by Leo Rauch (Detroit: Wayne State University Press, 1983), 87. For an online version of the text, see also: "The Philosophy of Spirit (Jena Lectures

1805–6)," Marxists.org, https://www.marxists.org/reference/archive/hegel/works/jl/cho1a.htm.

23. Ursula K. Heise, "The Hitchhiker's Guide to Ecocriticism," *PMLA* 121, no. 2 (March 2006), 507–8. Thoreau's saying comes from an essay called "Walking," in *The Great New Wilderness Debate* (Athens: University of Georgia Press, 1998), 31–41.

24. Jade Lindgaard, "Le grand bétonnage, une bombe climatique," *Mediapart*, July 27, 2015, https://www.mediapart.fr/journal/france/270715/le-grand-betonnage-une-bombe-climatique?onglet=full.

25. For more on these points, see Gary Snyder, *The Practice of the Wild* (1990; repr., Berkeley, Calif.: Counterpoint, 2010), 6, 9–14, as well as Jack Turner, "In Wildness Is the Preservation of the World," in *Deep Ecology for the Twenty-First Century*, ed. George Session (Boston: Shambhala, 1995), 334.

26. William Cronon, ed., "The Trouble with Wilderness, or Getting Back to the Wrong Nature," in *Uncommon Ground: Rethinking the Human Place in Nature*, ed. William Cronon (New York: W. W. Norton & Co., 1995), 88.

27. Cronon, "The Trouble with Wilderness," 88–89.

28. Mark Woods, "Wilderness," in *A Companion to Environmental Philosophy*, ed. Dale Jamieson (Malden, Mass.: Blackwell, 2001), 359.

29. Serge Moscovici, *Hommes domestiques et hommes sauvages* (Paris: UGE, 1974), 40–44; *De la nature: Pour penser l'ecologie* (Paris: Métaillé, 2002), 160.

30. Moscovici, *Hommes domestiques et hommes sauvages*, 160.

31. Maurice Merleau-Ponty, *Nature: Course Notes from the Collège de France*, trans. Robert Vallier (Evanston, Ill.: Northwestern University Press, 2003), 4.

32. [I have translated the French term *revers* by way of the two terms "dark side" and "counterlining" to grasp the variance of meanings inherent in the term. *Revers* can refer to both the *interior* or *lining* of a piece of cloth or to the *other side* of a coin, such as in the expression "le revers de la medaille," which demonstrates the polysemy of the term that Neyrat is striving for, as "le revers de la medaille" literally means "the other side of the coin" but is also employed to signify "on the other hand." "Dark side of counterlining" is thus my attempt to preserve this polysemy.]

33. Immanuel Kant, *Critique of Pure Reason*, vol. 2, trans. F. Max Müller (London: Macmillan and Co., 1881), 46–50.

34. For a critique of the epistemological privilege of the transcendental, see Steven Shaviro's *The Universe of Things* (Minneapolis: University of Minnesota Press, 2014), 3, 6.

13. THE UNCONSTRUCTABLE EARTH

1. Carolyn Merchant, *The Death of Nature: Women, Ecology, and the Scientific Revolution* (1980; repr., New York: Harper and Row, 1990), 293.

2. Alexander von Humboldt, *Cosmos: A Sketch of the Physical Description of the Universe*, vol. 1 (Baltimore: Johns Hopkins University Press, 1997), 39–40.

3. Cited by Paul Vidal de la Blache, "Le Principe de la géographie générale," *Annales de la géographie* 5, no. 20 (1896): 137. [For a recent English translation of the quote cited

by the author, see Carl Ritter, *Geographical Studies*, trans. William Leonhard Gage (1863; repr., London: Forgotten Books, 2015), 276.]

4. Vidal de la Blache, "Le Principe de la géographie générale," 129.

5. James Lovelock, "The UK Should Be Going Mad for Fracking," and "James Lovelock on Shale Gas and the Problem with 'greens,'" *The Guardian*, June 15, 2012.

6. Crutzen and Schwägerl assure us that they are inheritors of von Humboldt and his "world-organism" (see "Living in the Anthropocene," *Yale Environment 360*, January 24, 2011, https://e360.yale.edu/features/living_in_the_anthropocene_toward_a_new_global_ethos), but it's a very partial inheritance for authors who declare that we are nature and that we can modify it at will as if from the outside. Crutzen and Schwägerl's organicism stops at the gates of their geo-constructivism.

7. James Lovelock, *Gaia: A New Look at Life on Earth* (1979; repr., Oxford: Oxford University Press, 2016), 10.

8. Norbert Wiener, *The Human Use of Human Beings: Cybernetics and Society* (1950; repr., Boston: Houghton Mifflin, 1954), 26. However, the second cybernetics is established around the necessity for rethinking the singularity of the living (see the work of Franciso Varela, Evan Thompson, Eleanor Rosch, and Jon Kabat-Zinn, *The Embodied Mind: Cognitive Science and Human Experience* [1991; repr., Cambridge, Mass.: MIT Press, 2016], 35–39). For more on this theme see the postcybernetic analysis we propose in *Homo labyrinthus: humanisme, antihumanisme, posthumanisme* (Paris: Dehors, 2015), 137–50.

9. Here we are relying on the reading Clive Hamilton proposes regarding Lovelock's intellectual trajectory. Clive Hamilton, *Requiem for a Species* (Oxfordshire: Routledge, 2015), 147–52.

10. Lovelock, *Gaia*, 11.

11. Lynn Margulis, *Symbiotic Planet: A New Look at Evolution* (New York: Basic Books, 1999), 128.

12. Lovelock, *Gaia*, 103.

13. Wallace Broecker and Robert Kunzig, *Fixing Climate: What Past Climate Changes Reveal about the Current Threat—And How to Counter It* (New York: Three Books Publishing, 2008), 100.

14. Clive Hamilton, *Earthmasters: The Dawn of the Age of Climate Engineering* (New Haven, Conn.: Yale University Press, 2013), 37.

15. Concerning the term *unthinged*—unthingified—in Schelling, see Iain Hamilton Grant's *Philosophies of Nature after Schelling* (London: Continuum, 2006), 109.

16. Grant, *Philosophies of Nature after Schelling*, 137, 165.

17. Hannah Arendt, *The Human Condition*, 2nd ed. (Chicago: University of Chicago Press, 1998), 2.

18. Arendt, *Human Condition*, 10.

19. Arendt, 1.

20. Arendt, 2–3.

21. Edmund Husserl, "Foundational Investigations of the Phenomenological Origin of the Spatiality of Nature: The Originary Ark, the Earth, Does Not Move," in *Husserl*

at the Limits of Phenomenology: Including Texts by Edmund Husserl, Maurice Merleau-Ponty, trans. Leonard Lawlor and Bettina Bergo (Evanston, Ill.: Northwestern University Press, 2002), 118. Regarding the question of the Earth in Husserl as "other Nature" buried in the "depths of Cartesian nature," see Merleau-Ponty, *Nature: Course Notes from the Collège de France*, trans. Robert Vallier (Evanston, Ill.: Northwestern University Press, 2003).

22. Husserl, "Foundational Investigations," 118–19. [I've chosen to add both ways of translating what the author refers to as *sol* in French by way of *ground*, which is the term used in Fred Kersten's translation of the Husserl essay the author references (that appears in abovementioned publication of the Husserl text along with those by Merleau-Ponty) as well as the term *basis*, which is how Kersten translated the Husserlian concept in the other published version of the aforementioned essay found in *Husserl: Shorter Works* (Notre Dame, Ind.: University of Notre Dame Press, 1981), 222. Both translations are attempting to provide the extra semantic elements of originary foundation of the earth-body to which Husserl refers and that is central to his essay.]

23. Concerning this "world-without-us," see Eugene Thacker's *In the Dust of This Planet: Horror of Philosophy*, vol. 1 (Arlesford, UK: Zero Books, 2011), 5–7.

24. Relying on the work of Iain Hamilton Grant, Ben Woodard examines the reductionist tendency of "somatizing" the earth (i.e., reducing it to a primordial body that would be considered as the foundation of all thought). Ben Woodard, *On an Ungrounded Earth: Towards a New Geophilosophy* (Brooklyn, N.Y.: Punctum, 2013), 1–2, 5–12, regarding the metaphor of the Earth as island.

25. Clive Hamilton, Christophe Bonneuil, and François Gemenne, "Thinking the Anthropocene," in *The Anthropocene and the Global Environmental Crisis: Rethinking Modernity in a New Epoch*, ed. Clive Hamilton, Christophe Bonneuil, and François Gemenne (New York: Routledge, 2015), 5.

26. Eugene Thacker, *After Life* (Chicago: University of Chicago Press, 2010), xv, 268; Alan Weisman, *Homo disparitus* (the original, English title is *The World without Us*) (Paris: J'ai lu, 2007). Weisman imagines what would happen if humanity suddenly disappeared.

27. Jan Zalasiewicz, *The Earth after Us: What Legacy Will Humans Leave in the Rocks?* (New York: Oxford University Press, 2008).

28. Quentin Meillassoux, *After Finitude: An Essay on the Necessity of Contingency*, trans. Ray Brassier (London: Continuum, 2008), 26.

29. "Galileo's crime" consisted of "eliminating the living soul" from a world that was from then on mathematizable, giving way to a "de-natured environment" (Lewis Mumford, *The Pentagon of Power* [1964; repr., New York: Harcourt Brace Jovanovich, 1970], 51–68). The reflections relative to the "world-without-us" repeat through dramatizing, but without necessarily being aware of it, the modern moment.

30. "Pure science" here is a reference to Meillassoux; for Thacker, the access to a world-without-us (to the inhuman) is created by way of a demonology, an occult (of) science (see *In the Dust of This Planet*).

31. On the way in which "anteriority" must be "incorporated" into "substance," see Iain Hamilton Grant, "Mining Conditions: A Response to Harman," in *The Speculative*

Turn: Continental Materialism and Realism, ed. L. Bryant, G. Harman, and N. Srnicek (Melbourne: Re. Press, 2010), 44.

32. Zalasiewicz, *The Earth after Us*, 22–23.

33. Hans Blumenberg, *The Genesis of the Copernican World*, trans. Robert M. Wallace (Cambridge, Mass.: MIT Press, 1979), 679.

34. Benjamin Lazier uses the term *eccentricity* in his analyses of Blumenberg's work that have inspired us here. Benjamin Lazier, "Earthrise; Or the Globalization of the World Picture," *The American Historical Review* 116 (2011): 621–23.

35. Emmanuel Levinas, *Difficult Freedom: Essays on Judaism*, trans. Seán Hand (Baltimore, Md.: Johns Hopkins University Press, 1997), 233.

36. Levinas, *Difficult Freedom*, 231.

37. Blumenberg, *Genesis of the Copernican World*, 685.

38. Anselm Franke, "Earthrise and the Disappearance of the Outside," in *The Whole Earth: California and the Disappearance of the Outside*, ed. Diedrich Diederichsen and Anselm Franke (Berlin: Sternberg Press, 2013), 16.

39. Lewis Mumford, *Myth of the Machine: The Pentagon of Power* (1964; repr., New York: Harcourt Brace Jovanovich, 1970).

40. Bruno Latour, *Facing Gaia: Eight Lectures on the New Climatic Regime*, trans. Catherine Porter (Medford/Cambridge: Polity, 2017), 219, 281.

41. Latour, *Facing Gaia*, 244.

42. Latour, 40.

43. Latour, 289.

44. Latour, 284

45. Latour, 181.

46. Latour, 246. See as well the danger of becoming "foreigners in our own country" (277). In complete contrast to this statement, and in particular at the moment when we are writing these lines (November 2015), it seems ever more vital to leave a place to the Foreigner, to the foreigners who do not have, or no longer have, a country—to the refugees and those exiled.

47. See the analysis made by Jean-Baptiste Fressoz ("Cinécologie, episode 2—Écran carbone," *Débordements*, March 2015, http://www.debordements.fr/spip.php?article347), which we adhere to the extent that we are attempting here to dialectize the two films.

48. Nick Land appears to be at the controls of the blog *Outside In*, "Interstellar," *Outside In*, November 22, 2014, http://www.xenosystems.net/interstellar/. We should also take this opportunity to mention that Land is *also* the author of some remarkable and strange volumes of texts, *Fanged Noumena* that has already been cited: Nick Land, *Fanged Noumena: Collected Writings 1987–2007* (Falmouth/New York: Sequence Press, 2011).

49. We refer you here to Frédéric Neyrat's "Gravity ou comment revenir sur Terre," *Mediapart*, November 4, 2013.

50. For more on atopia as without-place, outside-place, see Frédéric Neyrat, *Atopias: Manifesto for a Radical Existentialism*, trans. Walt Hunter and Lindsay Turner (New York: Fordham University Press, 2017).

51. "Earth, Earth!" are the last two words of *Facing Gaia*. "Outside, here!" are our last two words on the dark side of the Earth.

CONCLUSION: WHAT IS TO BE UNMADE?

1. We should note in passing that leaving the Earth will not enable us to escape the entropy that reigns over the universe. The lone possible escape is not spatial, external; it is internal and temporal and consists of gathering up the universe within oneself.

2. [While we have followed the author's use of the French word *trajet* as a quasi-neologism because of its semantic rapport with the French words *sujet* (subject) and *objet* (object) and thus translated it as "traject" to maintain its rapport as neither subject nor object but perpetually in movement, it should nevertheless be noted that the typical definition of the French word *trajet* as journey or directional path contains in it this very notion of continuous movement, of a trajectory. For more on other similar interpretations of the translation of *trajet* as *traject* see the work of Paul Virilio and, in a much more in-depth manner, that of Augustin Berque. See in particular, Augustin Berque, *Écoumène: Introduction à l'étude des milieux humains* (2000; repr., Paris: Belin, 2015), and his conception of humanity as neither object nor subject but traject. See also Paul Virilio, *University of Disaster* (Cambridge: Polity, 2010).]

3. Georges Bataille, "L'économie à la mesure de l'univers," in *Œuvres complètes*, vol. 7 (Paris: Gallimard, 1976), 1–16. For more on Bataille and ecology, see Nigel Clark, *Inhuman Nature: Sociable Life on a Dynamic Planet* (Thousand Oaks, Calif.: *Theory, Culture, Society* and Sage Publications, 2010), 21–22.

4. Ivan Illich, *Tools for Conviviality* (New York: Harper & Row, 1973), 109.

5. André Gorz, *Ecologica*, trans. Chris Turner (London: Seagull Books, 2010), 7.

6. Gorz, *Ecologica*, 51–52.

7. Bruno Latour, *Facing Gaia: Eight Lectures on the New Climatic Regime*, trans. Catherine Porter (Cambridge: Polity, 2017), 247–48.

8. Latour, *Facing Gaia*, 255–77.

9. DATAR, "Des images de la France en l'an 2040," Des images de la France en l'an 2040: Rapport 2012 de la DATAR / DATAR (Délégation à l'aménagement du territoire et à l'attractivité régionale) (Paris: La Documentation française, 2012), 24, http://edd.ac -besancon.fr/wp-content/uploads/sites/12/2015/03/datar_des_images_de_la_france _en_l_an_2040.pdf.

10. Laetitia Van Eeckhout, "Manifestation au Burkina Faso contre les OGM de Monsanto," *Le Monde Afrique*, May 23, 2015.

11. Concerning a reading of contemporary political struggles either in terms of territories or antiextractivists, see Nichoal Haeringer, "Des ZAD, mais pour quoi faire?," *Le Monde*, December 14, 2014.

12. Nnimmo Bassey, "From South America to Africa, 'Capitalist' Solutions to Climate Change Seen as Path to Catastrophe," interview by Amy Goodman, Democracy Now!, December 10, 2014, https://www.democracynow.org/2014/12/10/from_south _america_to_africa_market (transcription). For more about the functioning of the

REDD mechanism, see Nnimmo Bassey, *To Cook a Continent: Destructive Extraction and the Climate Crisis in Africa* (Oxford: Pambazuka Press, 2012), 114–16.

13. Cited by Fred Magdoff and John Bellamy Foster in *What Every Environmentalist Needs to Know about Capitalism: A Citizen's Guide to Capitalism and the Environment* (New York: Monthly Review Press, 2011), 153.

14. André Gorz, *Écologie et liberté* (Paris: Galilée, 1977), 18, 92.

15. André Gorz, *Capitalism, Socialism, Ecology,* trans. Martin Chalmers (London: Verso, 2012). See p. 6–7 where Gorz refers to the Green "fundamentalist" Jurgen Dahl and his idea that in the face of the eventual collapse of capitalism "only poverty can save us" and could also thereby lead to the only means toward salvation as a true impetus for a change toward a sustainable social system citing the concept of an eventual "enforced renunciation" of wealth.

16. Serge Latouche, *Survivre au développement* (Paris: Mille et une nuits, 2004), 98–100.

17. See the work of Stéphane Haber, *Critique de l'antinaturalisme. Études sur Foucault, Butler, Habermas* (Paris: PUF, 2006).

18. Philippe Bihouix, "Du mythe de la croissance 'verte' à un monde post-croissance," in *Crime climatique stop!* (Paris: Seuil, 2015), 187–88, and *L'Âge des* Low Tech. *Vers une civilisation technologiquement soutenable* (Paris: Seuil, 2014), 69–72, 78, 95.

19. See Asher Miller and Rob Hopkins, "Climate after Growth: Why Environmentalists Must Embrace Post-growth Economics and Community Resilience," a report from the Post Carbon Institute, September 30, 2013, http://www.postcarbon.org/publications/climate-after-growth/.

20. Illich, *Tools for Conviviality,* 51.

21. Bill McKibben, "Recalculating the Climate Math," *New Republic,* September 22, 2016.

INDEX

of view of, 175, 177; terrestrial matter, 65;
and the wild, 161; within the inside, 127;
within the inside of the world, 127; of the
world, 19, 21; a world without, 183

panarchy, 80–81
Palmer, Michael, 179
People's Agreement of Cochabamba, 182
plan A, 31
plan B, 7, 30–31, 58, 123
plan C, 8, 180
political ecology, 6; anticonstructivist,
    10; and Black Lives Matter, 67; and
    division, 20; against geo-capitalism, 185;
    and geology, 179–80; and Latour, 102; as
    planetary ecology, 185; political ecology
    of separation, 182; and Stengers, 128;
    and the unconstructable, 21
post-environmentalist, 10, 13, 71, 83, 85, 88,
    93, 96, 100, 101, 106, 113, 118, 123, 124, 127
post-preservationist, 71, 85
pragmatist, 71, 83, 88, 105, 176
precautionary principle, 97, 101
Prigogine, Iya and Stengers, Isabelle, 11,
    73–75
principle of principle of ecology, 12, 146–48,
    153
Prometheus and Prometheanism, 10, 14, 33,
    35, 93, 98; Epiprometheus, 100, 119, 121;
    Promethean politics, 122

REDD (Reducing Emissions from Defor-
    estation and Forest Degradation), 182
Rees, Martin, 31, 32, 123
Reichardt, Kelly, 117
resilience, 10–11, 12, 15, 56–59, 74, 76–78,
    80, 81; and adaptation, 82; ecology of,
    83, 85, 86, 98, 119, 124, 181; and interior
    distance, 127; through separation, 183
Resilience Alliance, 76
Ritter, Carl, 165
Robock, Alan, 32
Romanticism, 106, 116–17
Rothenberg, David, 137
Ruddiman, William, 36–37

Sagan, Carl, 46–47
Schelling, F. W. J., 134, 156, 158, 159, 169
Schiller, Friedrich, 109
Schumpeter, Joseph, 79
Schwägerl, Christian, 64
separation: absence of, 139; abusive, 87; and
    continuity, 19; counterprinciple of, 14,
    153–54; and denial, 4; and hybridization,
    91, 94; illness of a lack of, 82; impossi-
    ble, 102; internal, 127; lack of, 137; and
    limit, 13; metabolization of, 15; political,
    185; power of, 103; and resilience, 183;
    and split, 149–52
Serres, Michel, 11, 92, 148
Shelley, Mary, 91, 98–99
Simon, L. Lewis and Maslin, Mark A., 38
Singer, Peter, 123
Sloterdijk, Peter, 124–25
slow dilapidation, 183
slow science, 128
Snyder, Gary, 107, 161
Space Age, 8, 40, 45, 47; end of the, 49; new
    space, 50–51; and the production of life,
    52, 122, 169
speculative realism, 148, 149
Spinoza, Baruch, 134, 136, 137–40; Society
    2.0, 143–44, 146, 159
Stengers, Isabelle, 12, 13, 120, 128, 147
stewardship and Earth stewardship, 2, 9,
    56–59, 61, 66, 86, 87, 181
Stiegler, Bernard, 98
Stockholm Resilience Center, 76
Stoermer, Eugene, 35, 37, 39
Saint-Simon, 60
synthetic biology, 52

technology: allotechnics and homeotech-
    nics, 124; as cosmotechnology, 125;
    and its conscripted apparatus, 125; and
    ecology of separation, 124; firefighter
    technology, 33; locking technologies
    and open technologies, 125; and magic,
    99; and precaution, 97; and preventive
    action, and unexpected consequences,
    100; *surexiste*, 120; and withdrawal, 96

Temple, Stanley, 53

Terminator (film), 121

terraforming, 8, 9, 45–47; and bacteria,
128, 144; domestic, 49; the Earth, 173;
and eco-modernists, 93; and life, 50–52,
82; terraform the planet, 89, 119, 144,
173

Thacker, Eugene, 171

Thoreau, Henry David, 160–62

traject, 20, 134, 170–73, 177, 180, 214n1, 223n2

transcendental narcissism: of the human
being, 134; and realism, 152

transhuman and transhumanism, 1, 5, 10,
72, 115, 122–24, 126, 174, 181

Tsing, Anna, 129

turbulence, 11, 12, 71; axiom of, 75, 79, 81,
83; mastery, 103; ontological axiom of
the paradigm of, 73; paradigm of, 121;
paradigm of the turbulence of, 119;
political economy of, 75–76, 81, 82; of
the world, 97

Turner, Frederick Jackson, 47

uncertainty, 11, 12; of chaotic systems,
74; and constructivism, 101; and
Fukushima, 103; and the precautionary
principle, 97; programmed uncertainty,
71; and resilience, 77

unconstructable, 19; base, 180; Earth, 177;
the furious ocean of the, 130; part of the
Earth, 21, 168; the, 163–164; *traject*, 171;

unconstructable share, 134; unconstructable trajectory, 20

Vidal de la Blache, Paul, 165

Von Bertalanffy, Ludwig, 42

Von Neumann, John, 27; Von Neumann
Syndrome, 95, 96, 103, 123

Wallace, Alfred Russel, 73

Weinberg, Alvin, 33

Weisman, Alan, 171

Whitehead, Alfred North, 11, 110, 134, 153,
155, 159, 163

wild: capacity of the Earth, 167; foundation,
177; as inner outside, 161; like the night
of time, 160; as the night of unfinished
natures, 163; re-wild, 53; as a spacing,
162; the, 17–18; and the transcendental
differential, 162; wild romanticism, 117

wilderness, 15, 47, 160; democratization of,
162; and wildness, 161

Wiener, Norbert, 166

Williams, Alex and Srnicek, Nick, 120, 122,
126, 127

Wood, Lowell, 56

Worster, Donald, 76, 107, 108

Zadists, 126

Zalasiewicz, Jan, 171, 173

Zarader, Marlène, 158

Žižek, Slavoj, 106, 107, 108

MEANING SYSTEMS

The Beginning of Heaven and Earth
Has No Name: Seven Days with
Second-Order Cybernetics. Edited by
Albert Müller and Karl H. Müller.
Translated by Elinor Rooks and
Michael Kasenbacher.
HEINZ VON FOERSTER

Cultural Techniques: Grids, Filters,
Doors, and Other Articulations of
the Real. Translated by Geoffrey
Winthrop-Young.
BERNHARD SIEGERT

Interdependence: Biology and Beyond.
KRITI SHARMA

Earth, Life, and System: Evolution
and Ecology on a Gaian Planet.
BRUCE CLARKE (ED.)

Upside-Down Gods and Gregory
Bateson's World of Difference.
PETER HARRIES-JONES

The Technological Introject: Friedrich
Kittler between Implementation and
the Incalcuable.
JEFFREY CHAMPLIN AND
ANTJE PFANNKUCHEN (EDS.)

Google Me: One-Click Democracy.
Translated by Michael Syrotinski.
BARBARA CASSIN

The Unconstructable Earth: An
Ecology of Separation. Translated
by Drew S. Burk.
FRÉDÉRIC NEYRAT

Printed and bound by CPI Group (UK) Ltd, Croydon, CR0 4YY

27/10/2024

14580327-0001